SpringerBriefs in Applied Sciences and Technology

Automotive Engineering : Simulation and Validation Methods

Series Editors

Anton Fuchs, Virtual Vehicle Research GmbH, Graz, Austria

Hermann Steffan, Graz University of Technology, Graz, Austria

Jost Bernasch, Virtual Vehicle Research GmbH, Graz, Austria

Daniel Watzenig, Virtual Vehicle Research GmbH, Graz University of Technology, Graz, Austria

More information about this subseries at http://www.springer.com/series/11667

Anton Fuchs · Bernhard Brandstätter
Editors

Future Interior Concepts

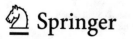 Springer

Editors
Anton Fuchs
Virtual Vehicle Research GmbH
Graz, Austria

Bernhard Brandstätter
Virtual Vehicle Research GmbH
Graz, Austria

ISSN 2191-530X   ISSN 2191-5318   (electronic)
SpringerBriefs in Applied Sciences and Technology
ISSN 2570-4028   ISSN 2570-4036   (electronic)
Automotive Engineering : Simulation and Validation Methods
ISBN 978-3-030-51043-5   ISBN 978-3-030-51044-2   (eBook)
https://doi.org/10.1007/978-3-030-51044-2

This Springer imprint is published by the registered company Springer Nature Switzerland AG
The registered company address is: Gewerbestrasse 11, 6330 Cham, Switzerland

Preface

Today's vehicle occupants place increasingly challenging demands on the interior of the vehicle cabin. Comfort—with all its peculiarities, such as thermal, acoustic, haptic, olfactory, safety and security and trust in new technologies and how to interact with the car—is for sure the central momentum for car manufacturers to differentiate, since cabin comfort is the first impression when buying a car and is most disturbing and annoying, when expectations and needs are not met.

Thus, future interior design requires to deal with a large number of technical fields, influencing factors, perception and psychology, covering, e.g., digitalization and human–machine interfaces, communication, smart surfaces and sensors, passenger monitoring, new vehicle safety aspects, sustainable materials, etc. A comprehensive investigation of the interior concepts also comprises highly subjective parameters such as well-being, the effect of colors or sounds or trust.

New requirements for multi-functional and adaptive spaces that "follow" the driver and passenger foster innovations. When a driver changes seat position in future highly automated driving mode, it requires heating ventilation and air conditioning (HVAC) systems, restraint and safety systems, in-vehicle entertainment and communication, operation and control systems, etc. to react and adapt as well.

The "comfortable vehicle" includes a wide range of topics and domains from thermal comfort, noise, vibration, and harshness (NVH) and driver assistance systems and requires innovative thoughts from different directions.

Supported by improving comfort in our daily work and leisure time in fields such as ambient assisted living, building automation, or smartphone technology, the expectations concerning comfort have increased as well and gained specific importance in mobility. Automated driving—and autonomous driving—will make the passenger compartment even more a place of relaxing transport, communication, and efficiently working in the future.

Defining "comfort" typically includes rather subjective impressions. It is mostly associated with the feeling of relaxation and well-being. Developing the vehicle comfort for a large group of customers and future passengers means to design a well-balanced sum of individual aspects and measures—each of them with individual weight but all relevant (a car might also perceive comfort needs of its individual drivers and passengers and may adapt to it).

What will be the value of a perfect thermal comfort in the passenger compartment, when it is distorted by an annoying fan noise or an undesirable smell of the HVAC system? And so would the insufficient ease of use of a human–machine interface (HMI) reduce the comfort impression of advanced driver assistance systems (ADAS).

This book will highlight selected fields and development methods for future interior concepts in automotive industry, VIRTUAL VEHICLE, and institutions contributing to this book are researching. It comprises a chapter on seat–human interaction and perception, vibro-acoustic metamaterials for improved interior NVH performance, active sound control, psychological HMI aspects, and investigations for the vehicle interior as well as a comprehensive thermal management and comfort consideration.

Graz, Austria   Bernhard Brandstätter
April 2020   Anton Fuchs

Contents

Seat-Human Interaction and Perception: A Multi-factorial-Problem

M. Wegner, C. Reuter, F. Fitzen, S. Anjani, and P. Vink

Abstract This study investigates the tactile perceived seat-human interaction of four types of BMW 5-series seats with the same foam properties and contours but different seat cover and seat suspension properties; 38 healthy subjects participated in an experiment rating and ranking the tactile perceived properties of the seats while blindfolded. A discomfort test, a seat characterizing rating on a scale of word pairs, and the overall experience of the seats were examined in four different sitting positions. The results of the experiment were related with the outcome of an objective measurement method: a pressure measurement mat and the measurement tool of Wegner et al. [19]. The study showed that the perception of the surface while interacting with the seat is independent from the sitting position. In contrast, the perception of the hardness and the elasticity of the seat is position-dependent. The results of the seat characterization are in line with the results of the measurement tool of Wegner et al. [19]. Further research is needed to investigate the mutual interdependence of the various measurement points of the measurement tool and to improve the prediction accuracy of the seat characteristics.

Keywords Pressure measurements · Shear force · Discomfort · Seat perception

1 Introduction

Most individuals, and particularly those with sedentary jobs, sit for nearly ten hours each work day and eight hours during their own, independent leisure time [14]. Typically, as long as the individual feels comfortable and supported, the seat on which an individual is seated is of little importance. Regardless of what seat and

M. Wegner (✉) · P. Vink
Faculty of Industrial Design Engineering, Delft University of Technology, Landbergstraat 15, 2628 CE Delft, The Netherlands
e-mail: Maximilian.Wegner@bmw.de

M. Wegner · C. Reuter · F. Fitzen · S. Anjani
BMW GROUP Research and Innovation Centre (FIZ), BMW Group, Knorrstraße 147, 80937 Munich, Germany

© The Author(s), under exclusive license to Springer Nature Switzerland AG 2021
A. Fuchs and B. Brandstätter (eds.), *Future Interior Concepts*,
Automotive Engineering : Simulation and Validation Methods,
https://doi.org/10.1007/978-3-030-51044-2_1

1

what position a person takes, the seat or chair should allow to vary and shift the posture easily. In this context Sammonds et al. [16] showed that movements and seat fidgets correlate with the discomfort rating of a seat. The micro and macro movements rise over the duration of time as well as the poor subjective discomfort ratings.

The development of seats for automobiles that allow passengers to move and switch to various positions from sitting through to lying is crucial to the automotive industry. This could become even more important in autonomous driving cars as more seat positions will be possible when there is no driving task. For an individual to be comfortable in the car, a car seat must support the passenger in a dynamic driving situation but moreover provide enough space for postural changes in various loading situations. Hence, it should be considered to change loading of the area of the seat being in contact with the passenger as well as the interaction area including various sensitivity areas. A study by Vink and Lips [18] proved that the pressure sensitivity of the area touching the shoulder and the area touching the front of the cushion close to the knees is significantly higher than all other body areas in contact with the seat. Furthermore, some parts of the body need more support than others. Biedermann and Guttmann [1] claimed, inter alia, that the natural physiological curve of the spine should be supported in the lumbar area. There are more influencing factors [20] making the discomfort and comfort perception of an automotive seat a multi-factorial problem with contributions occurring from effects of the seat layout including the foam properties, the contour, the cover properties, and the dynamic environment as well as effects on the human senses including the sitting, position, the sitting duration, pressure, shear force, and blood flow.

Most studies focus on the driver position and on the discomfort ratings of seat contours and seat foams relating the findings to pressure parameters (e.g. [9, 12, 21]). However, the multi-factorial problem is often reduced to a mono-problem, not taking the seat cover and other seat components into account. Most studies neglect to address other interactions parameters of the human senses than pressure. Mansfield et al. [13] investigated the extent of which foam properties affect the discomfort rating. For his study he removed the seat cover in order to enable the foam being in direct contact with the subject's clothing. Also, Hiemstra-van Mastrigt [11] compare the foam hardness of two train seats and checked the effect on comfort experience. Zenk et al. [21] used various foams to evoke different pressure distributions and thus different discomfort ratings. In reference to this approach an ideal pressure distribution was developed and after validated in a long-term rating. The results represent that there is a link between the cushion, the discomfort rating, and the pressure distribution of the cushion. Notably, the correlation between the backrest was not significant. Both, Mansfield et al. [13] and Zenk et al. [21] excluded the surface, cover properties of the seat, and the interaction of the seat components.

In contrast, Zuo et al. [22] revealed that the sensory properties of materials are relevant for the interaction between users and should be considered in the course of the material selection process. Regarding the gathered information he developed a method for an intelligent choice of materials based on holistic perceptional information of different materials. Likewise, Wegner et al. [20] showed that the seat cover material has fundamental influence on the perception and the characterization of a seat. The study compares two seats with the same contour and the same foam properties but with different cover materials.

With reference to the human mechanoreceptors explained by Schmidt and Thews [17], not only the pressure is an important tactile sensor but also the shear and the elongation have to be taken into account. Chow and Odell [2] linked the pressure to shear stress stating that interface shear force significantly affects the pressure distribution. Based on simulative results Grujicic et al. [6] correlated a higher cover friction to higher shear forces. Also, Goossens and Snijders [5] showed that the shear force could be reduced by changing the seat position and seat angles on the one hand. On the other hand, Goossens [7] presented that the shear force can be reduced by using the right cushion material, a LiquiCell cushion. Thus, not only the ideal seat angle [10], seat pan angle of 10° and backrest angle of 120° is important but additionally the angle position in combination with the applied seat components.

In this study the seat perception is considered as a multi-factorial problem including various seat components as well as the seat-human interaction parameters: pressure, elongation and shear force [17]. The aim for this study is to investigate how occupants rate and perceive seat characteristics and discomfort of car seats with equal foam properties and contours but different cover properties and seat suspensions in various loading states. Next, the study investigates whether the objective measurement methods with the pressure measurement mat and the measurement tool of Wegner et al. [19] sufficiently explain the seat ratings.

2 Methods

In this section the study approach: the scope of participants, the seats used for the study, the procedure of the study, and the statistical analysis are presented. The description of the procedure also includes the presentation of two objective seat measurement methods: first, the pressure measurement mat and second the seat measurement with the measurement tool of Wegner et al. [19].

2.1 Participants

38 subjects, 17 males and 21 females, participated in the experiment. The mean body height of the participants was 1.69 m (1.53–1.86 m) with a mean body weight of 66.2 kg (48–98 kg). On the torso, the participants either whore t-shirts (60%), pullovers (16%), long sleeve t-shirts (11%), polo shirts (8%), or dresses (5%); on the bottom either jeans (55%), cloth pants (40%), or leggings (5%).

2.2 Seats

Four BMW 5-series seats are used in this study. The standard contour of the seats was used, which is not distinctive. The seat layout was kept simple, consisting of a seat frame, foam, heating mat, and cover. All seats are produced and assembled in the same factory on the same day, and during a similar period fulfilling all specified requirements of the manufacturer, especially the foam hardness which is measured in kPa. One seat, defined as the reference seat, is without any modification (*seat 1*). *Seat 1* is a leather seat with a specified foam hardness of 6 kPa in the main surface of the cushion and 10 kPa in the bolsters. The backrest has a foam hardness specification of 5 kPa in the main surface and 8 kPa in the bolsters. Compared to the reference seat, each seat differs in one parameter: One seat has an Alcantara cover instead of leather (*seat 2*), another seat (*seat 3*) has a looser cover tension, and the last seat has a metal plate installed instead of the original seat suspension (*seat 4*).

2.3 Setup

The four seats are mounted next to each other on a base plate (Fig. 1). The plate has a footrest following the geometric specifications of the BMW 5-series. All seats have an electrical seat adjustment which allows to adjust all seats equally to four different positions (Table 1). Position 1 is the driving position, containing the required seat angles for development of the seat and safety requirements. Position 2 and 3 have a flat cushion angle with the difference that the backrest angle in Position 3 is more horizontal than in Position 2. Position 3 and 4 have the same γ-angle but Position

Fig. 1 The figure illustrates the setup of the study with all four seats in a row from left to right: reference seat (*seat 1*); Alcantara seat (*seat 2*); loose cover tension (*seat 3*), and the seat with the metal plate instead of the seat suspension (*seat 4*)

Table 1 Illustration of the four adjusted seat angles for the cushion and the backrest

	α	β	γ
Position 1	14°	20°	96°
Position 2	3°	40°	127°
Position 3	3°	55°	142°
Position 4	18°	70	142°

4 has a higher cushion (α) and backrest (β) angle. The reason for these position changes was to create changes in comfort perception and pressure distribution as by the variation of the angles the weight of the body loads the cushion and backrest differently.

2.4 Procedure

2.4.1 Seat Evaluation

For gathering anthropometrics data, an anthropometric chair was used. Data regarding sitting height, hip width, buttock-popliteal length etc. were recorded using the procedure described by Molenbroek et al. [15]. During the recording, which took several minutes, each participant was informed about the procedure and the questionnaire but did not get any information regarding the setup and the differences of the seats. The participants were blindfolded wearing an eye mask during the entire experiment in order to exclude visual impressions. Only one participant at a time was going through the procedure. Once all tests were completed the next participant started. This way the participants could not exchange any information prior to the test. The study began with the participants discomfort rating of all four seats in Position 1. The order in which the participants rated the seats was changed for all tests systematically. The participants were not allowed to touch the seat surface. After sitting three minutes in each seat, the participants rated the discomfort of the seats through a Local Postural Discomfort (LPD) body map and a discomfort score from zero (no discomfort) to six (very heavy discomfort). Afterwards, for each seat and each participant a pressure measurement was conducted in Position 1. Regarding the pressure analysis the cushion is divided in three groups shown in Fig. 2: *buttock Group*, *front Group* and *side Group*. The backrest is cumulated into another group, called *back Group*. For every participant the recorded frames per each group were merged and the *average pressure*, *peak pressure*, and *contact area* calculated. The mean value and the standard deviation for the *average pressure*, *peak pressure*, and *contact area* over all 38 participants and for each seat were determined.

Next, the participants had to rate with words each seat in all of the four positions (Table 1). Three pairs of words given for them to describe the cushion and the backrest: *soft-hard*, *elastic-stiff*, and *slippery-abrasive*. The word pairs are shown

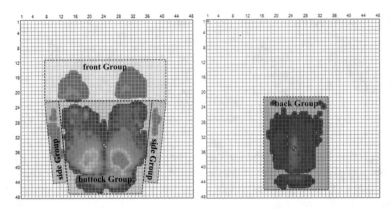

Fig. 2 Considered areas of pressure defined in three groups for the cushion and one group for the backrest

on a Likert Scale (1–7). Ratings of 1, 2, or 3 represent a tendency to a soft, elastic, and slippery characterization whereas ratings of 5, 6, or 7 have a tendency to a hard, stiff, or abrasive characterization. A rating of 4 demonstrates a neutral rating without any tendency to one of the extremes. After rating all four seats the participants were asked to rank the seats from their favorite to their least favorite seat.

2.4.2 Measuring the Seats with a Measurement Tool

After the test was conducted the seats were analyzed with the measurement tool of Wegner et al. [19]. The measurement points (Fig. 3) for the backrest are the shoulder (1), the lumbar area (2), and the bolster of the backrest (3). The measurement points for the cushion are at the area of the ischial tuberosity (4), the front of the cushion (5), and the bolster of the cushion (6).

The measurement procedure for each measurement point includes four cycles, three pre-cycles, and one measurement cycle (following the guidelines in DIN 53579 [3] and DIN EN ISO 3386-1 [4]. The measurement cycle has four phases (see Fig. 5). During the first phase (①) the stamp loads the seat with a velocity of 100 mm/min until 100 N is reached. During the second phase (②) the stamp remains in the position for 30 s. Hereafter, the machine adjusts during the third phase (③) the force again up to 100 N and moves the Seat 5 mm in lateral direction relative to the stamp and remains 15 s in this position. The fourth phase (④) is the relief phase (300 mm/min).

During this measurement procedure the sensors of the stamp (five pressure sensors and fore elongations sensors, Fig. 4) record constantly the properties of the seats in each measurement point. The stamp has a silicon surface simulating the human skin. Figure 5 shows an example of the recorded data for a seat in one of the six measurement points. The first plot shows the recordings of the force and indentation. This plot includes the division into the four measurement phases (①–④). The second plot shows the recordings of the five pressure sensors (1–5). Last, the third plot

Fig. 3 Illustration of the measurement points

Fig. 4 Detailed illustration of the stamp. Pressure sensor are named from 1 to 5 and elongations sensors from I to IV

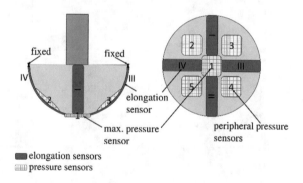

exposes the recording of the elongation sensors (I–IV). Based on these plots the following parameters for pressure and elongation are calculated.

Pressure: The (1) *first touch pressure* is defined as the pressure information of pressure sensor 1 after 5 mm indentation (empirical defined value of BMW internal Comfort Experts). The (2) *maximum pressure* has been defined as the value of pressure sensor 1 when a force of 100 N is reached. The (3) *linear pressure* identifies the shift from a linear rise of the pressure to an exponential rise of pressure based on the values of sensor 1 (first phase ①). The (4) *pressure distribution* is defined as the average pressure of the peripheral pressure sensors (sensor 2–4, Fig. 4) in phase two (②). The *maximum pressure* and the *linear pressure* are linked to the indentation information ((7) *linear indentation*, (8) *maximum indentation*).

Elongation: While loading (first phase ①), the elongation of each of the four sensors is recorded. The information of sensor I, II, III, and IV is summed to an

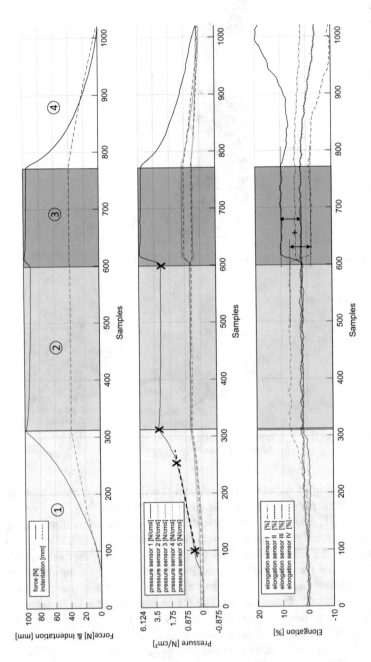

Fig. 5 The top diagram illustrates the force and indentation, the middle diagram presents the sensor data of the five pressure sensors, and diagram at the bottom presents the four elnongations sensor data

overall elongation (5) *elongation while loading the seat*. The information of the elongation sensor III and IV in phase three (③) enables to calculate the change of the elongation while applying a shear stress (moving the seat relative to the stamp in the direction of sensor III and IV). The change of elongation sensor III and IV is identified by calculating the difference between phase two (②) and phase three (③) of each sensor. Both values of sensor III and IV are summed up to an overall (6) *elongation due to the lateral movement*.

For a better comparability of the seats the (2) *maximum pressure* is normalized with the (8) *maximum indentation* and the (3) *linear pressure* is normalized with the (7) *linear indentation*. The (5) *elongation while loading the seat* and the (6) *elongation due to the lateral movement* are both normalized with a factor consisting the multiplication of the (2) *maximum pressure* and the *friction coefficient*. The (1) *first touch pressure* and the (4) *pressure distribution* are not normalized.

2.4.3 Determination of the Friction Coefficients

For an adequate comparison of both seat cover materials (leather and Alcantara) static and dynamic friction coefficient tests are conducted. The following material pairs are tested: *leather–silicon, Alcantara–silicon, leather–jeans, Alcantara–jeans*. By testing the friction coefficients of leather and Alcantara in combination with silicon and jeans a conclusion on the differences between silicon and jeans material could be made.

2.5 Statistical Analysis

The data of the word pair ratings were analyzed using a statistical analysis software program (IBM SPSS Statistics 25). The Friedman's Test was used to determine whether the participants detect differences in the perception of the four seats. The analysis was separately done for the cushions and the backrests ($\alpha < 0.05$) regarding their sitting position. If the results of the Friedman's Test are significant a post hoc analysis with a Wilcoxon signed-rank test is conducted for all six seat combinations (e.g., *seat 1–seat 2* or *seat 2–seat 4*). The six seat combinations are treated as six separate and unrelated observations, therefore, the Bonferroni correction is not applied, and the statistical significance is set to $\alpha < 0.05$.

3 Results

In the following section the results of the discomfort ratings are presented first. After this the descriptive results of the word pair ratings in each of the four positions is presented. Furthermore, the results of the Friedman's Test and the Wilcoxon signed-rank test are presented. Eventually, the last part illustrates the results of the pressure measurements and the analysis of the four seats with the measurement tool of Wegner et al. [20] as well as the results of the friction coefficient measurements.

3.1 Subjective Perception of the Seats

3.1.1 Discomfort Rating

Table 2 shows the results of the discomfort rating of the four seats. Ratings higher than 0 indicate discomfort. Regions with more than two complaints ($N > 2$) are bold. Regarding the cushion most participants have discomfort complaints in the second seat, the Alcantara seat. Discomfort appears to be large for the rear bolster region (H1 and H2) and in the front of the main surface (G1 and G2).

Regarding the backrest the reference seat (*seat 1*) has only one noticeable complaint; four participants mentioned discomfort in the upper back. The modified seats have all discomfort in the outer shoulder area (D1, D2), whereas *seat 3* has the most noticeable discomfort. For the same seat also in the backrest bolsters (E1, E2) noticeable discomfort complaints were issued. *Seat 4* (seat without seat suspension) has also noticeable discomfort complaint in the lumbar area (B2).

Participants who mentioned discomfort it was predominantly high in more than two areas for one seat. Nevertheless, the Alcantara seat (*seat 2*) has most discomfort in the cushion area and the seat with the loose cover tension (*seat 3*) as well as the seat with a plate instead of the seat suspension (*seat 4*) have high discomfort in the backrest areas.

3.1.2 Word Pair Rating

Descriptive: Figure 6 gives a descriptive overview of the seat and position characteristics. The orange circle represents the neutral rating (Likert Scale rating of 4). Every characteristic which is rated hard, stiff, or abrasive lies outside the circle and the characteristics soft, elastic, and slippery lie inside the circle. Figure 6 illustrates that *seat 3* in Position 1, the driving position, is rated as the softest and the most elastic seat. In contrast, all other seats are rated stiffer for the backrest as well as for the cushion. *Seat 2* is rated as the most abrasive seat especially for the backrest. The seat rated the hardest regarding the cushion and the backrest is *seat 4*. As for Position 2 the abrasive surface of *seat 2* appears dominant for the participants. Furthermore, the hardness of the backrest of *seat 4* is dominant. Overall, in Position 2 all other ratings of the characteristics move closer to the neutral rating. In Position 3 the abrasive surface of *seat 2* is still dominant to the participants. Other than that, all seats in Positions 3 are rated harder and stiffer for the backrest than in Position 1 and 2. As opposed to Position 2 and 3, in which most characteristics for the four seats were

Table 2 Results of the discomfort rating with a local postural discomfort (LPD) body map for all four seats

		Seat 1 reference (leather)		Seat 2 Alcantara		Seat 3 loose cover (leather)		Seat 4 plate (leather)	
		N	Ø-rating	N	Ø-rating	N	Ø-rating	N	Ø-rating
Backrest	A	0	0	0	0	2	2	2	1.5
	B1	4	2.5	1	1	2	1.5	2	3
	B2	2	2	1	1	3	3	4	2.25
	C	0	0	0	0	0	0	0	0
	D1	2	1	3	1.33	5	2.8	3	1.33
	D2	2	1	3	1.33	5	2.8	3	1.33
	E1	1	4	1	3	3	2.67	1	4
	E2	1	4	1	3	3	2.67	1	4
Cushion	F1	2	1	0	0	0	0	1	3
	F2	2	1	0	0	0	0	1	3
	G1	2	1.5	3	1.33	1	2	2	2
	G2	2	1.5	3	1.33	1	2	2	2
	H1	1	1	5	2	1	1	1	1
	H2	1	1	5	2	1	1	1	1
	I1	0	0	0	0	1	1	1	1
	I2	0	0	0	0	1	1	1	1

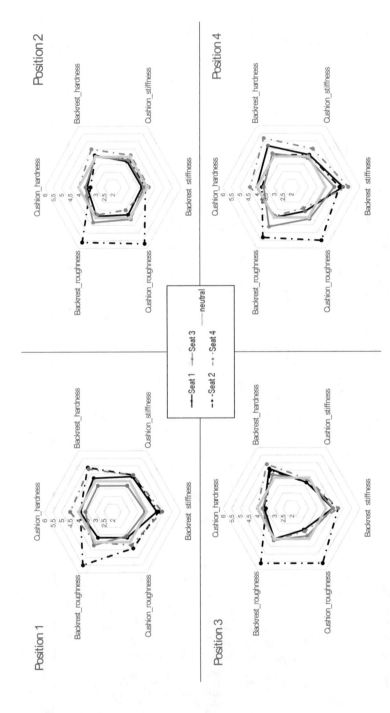

Fig. 6 Illustration of the rated seat characteristics for the Position 1, Position 2, Position 3, Position 4

rated similarly, the ratings and the characterizations in Position 4 are different for all four seat. For Position 4 *seat 1* is rated slippery in the cushion and hard and stiff in the backrest. *Seat 2* is rated abrasive in cushion and backrest and stiff in backrest. *Seat 3* is rated abrasive and stiff in backrest and *seat 4* is rated hard and stiff in the cushion and backrest.

A detailed listing of the means and the standard deviations for each seat in each position is presented in the appendix. All in all, *seat 1* received a rather neutral rating but has in some positions (Position 1 and Position 3) slippery characteristics. *Seat 2* is according to the ratings in each position the most abrasive seat regarding the cushion and the backrest and is also rated the softest either for the cushion or the backrest in each position except for Position 1. *Seat 3* is rated as the softest seat in Position 1 for the backrest and cushion and for Position 2, 3, and 4 as the softest either for cushion or the backrest. *Seat 3* is moreover rated the most elastic seat. *Seat 4* is rated the hardest seat regarding cushion and the backrest and also the most stiff and most slippery for the cushion and the backrest.

Statistical analysis:

Position 1: For the cushion the results of the Friedman's Test indicated a significance for all three word pairs: *soft–hard* ($\chi2(3) = 12.77, p = 0.005$), *elastic–stiff* ($\chi2(3) = 8.21, p = 0.042$) and *slippery–abrasive* ($\chi2(3) = 32.55, p = 0.001$). Each word pair is used to differentiate between the four seats. Also for the backrest the differentiation of the four seats is for all three word pairs significant: *soft–hard* ($\chi2(3) = 20.61, p = 0.001$), *elastic–stiff* ($\chi2(3) = 19.22, p = 0.001$) and *slippery–abrasive* ($\chi2(3) = 30.68, p = 0.001$).

Table 3 illustrates the results of the post hoc analysis using the Wilcoxon signed-rank test ($\alpha < 0.05$). The Wilcoxon test presents for the word pair *soft–hard* significances in the cushion for the following seat pairings: *seat 1–seat 4, seat 2–seat 4, seat 3–seat 4*. Thus, it is clear that *seat 4* (metal plate instead of a seat suspension) has the highest load on the cushion in Position 1, because *seat 4* is present in each word pair that shows significance. The backrest shows significances for the same set of seat parings and furthermore for seat pairing: *seat 1–seat 3* (reference seat and the seat with a loose cover tension). For the word pair *elastic–stiff* the results of the Wilcoxon signed-rank test present the same significant seat pairings for cushion and backrest: *seat 1–seat 3, seat 2–seat 3*, and *seat 4–seat 3*. In this case each seat pairing contains *seat 3* with the loose cover tension. For the word pair *slippery–abrasive* the significant seat pairings of the Wilcoxon signed-rank test are the same also for the cushion and backrest: *seat 1–seat 2, seat 2–seat 3, seat 2–seat 4*. In this case the *seat 2* with the Alcantara cover is in each of the pairings present.

Position 2: For Position 2 the Friedman's Test indicates significant differences of the seat cushion for the word pairs *soft–hard* ($\chi2(3) = 8.80, p = 0.032$) and *slippery–abrasive* ($\chi2(3) = 36.14, p = 0.001$). For the backrest the word pair *slippery–abrasive* ($\chi2(3) = 41.34, p = 0.001$) indicates significance in differentiation.

Table 4 demonstrates the results of the Wilcoxon signed-rank test for the cushion and the backrest in each seat pairing combination. Concerning the word pair *soft–hard* the Wilcoxon singed-rank test points out that there are significant differences

Table 3 Results of the Wilcoxon sign-rank test for Position 1

Position 1		Seat 1–seat 2	Seat 1–seat 3	Seat 1–seat 4	Seat 2–seat 3	Seat 2–seat 4	Seat 3–seat 4
Soft–hard	*Cushion*						
	Z	−0.365	−1.222	−2.129	−0.802	−2.196	−3.412
	P	0.715	0.222	**0.033**	0.423	**0.028**	**0.001**
	Backrest						
	Z	−1.232	−2.202	−2.623	−1.020	−2.437	−3.868
	P	0.218	**0.028**	**0.008**	0.308	**0.015**	**0.000**
Elastic–stiff	*Cushion*						
	Z	−0.440	−2.525	−0.243	−2.224	−0.058	−2.239
	P	0.66	**0.012**	0.808	**0.026**	0.954	**0.025**
	Backrest						
	Z	−0.208	−2.967	−0.922	−2.697	−1.111	−3.378
	P	0.835	**0.003**	0.356	**0.007**	0.266	**0.001**
Slippery–abrasive	*Cushion*						
	Z	−4.382	−1.281	−0.037	−3.713	−4.274	−1.251
	P	**0.000**	0.200	0.970	**0.000**	**0.000**	0.211
	Backrest						
	Z	−4.060	−1.457	−0.726	−3.613	−4.030	−0.822
	P	**0.000**	0.145	0.468	**0.000**	**0.000**	0.411

Table 4 Results of the Wilcoxon sign-rank test for Position 2

Position 2		Seat 1–seat 2	Seat 1–seat 3	Seat 1–seat 4	Seat 2–seat 3	Seat 2–seat 4	Seat 3–seat 4
Soft–hard	*Cushion*						
	Z	−0.741	−0.502	−2.210	−1.230	−2.413	−1.749
	P	0.458	0.615	**0.027**	0.219	**0.016**	0.080
Slippery–abrasive	*Cushion*						
	Z	−4.389	−0.962	−1.312	−3.940	−4.455	−2.064
	P	**0.000**	0.336	0.189	**0.000**	**0.000**	**0.039**
	Backrest						
	Z	−4.360	−1.852	−0.030	−4.094	−4.491	−1.715
	P	**0.000**	0.064	0.976	**0.000**	**0.000**	0.086

for the seat pairings: *seat 1–seat 4* and *seat 2–seat 4*. Both seat pairings include *seat 4*. The Wilcoxon signed-rank test results referring to the word pair *slippery–abrasive* have the same significant seat pairings for the cushion and backrest: *seat 1–seat 2*, *seat 2–seat 3*, *seat 2–seat 4*. All seat combinations contain the *seat 2*.

Table 5 Results of the Wilcoxon sign-rank test for Position 3

Position 3		Seat 1–seat 2	Seat 1–seat 3	Seat 1–seat 4	Seat 2–seat 3	Seat 2–seat 4	Seat 3–seat 4
Soft–hard	*Backrest*						
	Z	−0.751	−1.727	−0.931	−1.501	−1.437	−2.213
	P	0.453	0.084	0.352	0.133	0.151	**0.027**
Slippery–abrasive	*Cushion*						
	Z	−5.100	−1.532	−0.787	−4.626	−5.049	−1.207
	P	**0.000**	0.125	0.431	**0.000**	**0.000**	0.228
	Backrest						
	Z	−4.511	−0.546	−0.559	−4.122	−4.448	−0.222
	P	**0.000**	0.585	0.576	**0.000**	**0.000**	0.824

Position 3: The results of the Friedman's Test are significant for the word pair *slippery–abrasive* for the cushion ($\chi 2(3) = 56.01$, $p = 0.001$) as well as for the backrest ($\chi 2(3) = 36.72$, $p = 0.001$). The word pair *soft–hard* ($\chi 2(3) = 10.07$, $p = 0.018$) is only significant for the backrest.

Table 5 exposes for the backrest regarding the word pair *soft–hard* only one significant seat pairing: *seat 3–seat 4*. With reference to the word pair *slippery–abrasive* the cushion as well as the backrest have the same seat pairings with significant results of the Wilcoxon signed-rank test. The significant seat pairings are: *seat 1–seat 2, seat 2–seat 3, seat 2–seat 4*. In all seat pairings *seat 2* with the Alcantara cover is present.

Position 4: The Friedman's Test is significant for the backrest for all three word pairs: *soft–hard* ($\chi 2(3) = 21.54$, $p = 0.001$), *elastic–stiff* ($\chi 2(3) = 16.22$, $p = 0.001$), and *slippery–abrasive* ($\chi 2(3) = 29.25$, $p = 0.001$). As to the cushion the word pairs *soft–hard* ($\chi 2(3) = 13.19$, $p = 0.004$) and *slippery–abrasive* ($\chi 2(3) = 44.64$, $p = 0.001$) are significant (Table 6).

The Wilcoxon singed-rank test lays out, that in respect to the cushion and the word pair *soft–hard* the seat pairings *seat 2–seat 4* and *seat 3–seat 4* are significant for differentiation. *Seat 1* is not included in the differentiation of hardness (word pair *soft–hard*). Thus, for Position 4 the differentiation of the hardness for the cushion is perceived between the *seat 4* with a plate instead of a seat suspension and *seat 3* with loose cover tension or *seat 2* with an Alcantara cover. As for the word pair *slippery– abrasive* all seat pairings are significant for differentiation, except seat pairing *seat 1– seat 4*, which is the reference seat compared to the seat without a seat suspension. The backrests can be differentiated regarding the word pair *soft–hard* with the significant seat pairings: *seat 1–seat 3, seat 2–seat 4* and *seat 3–seat 4*; the word pair *elastic–stiff* with the significant word pairings: *seat 1–seat 3* and *seat 3–seat 4*; and the word pair *slippery–abrasive* with the seat pairing: *seat 1–seat 2, seat 1–seat 3, seat 2–seat 3, seat 2–seat 4* and *seat 3–seat 4*. The results for the cushion do not include the seat pairing *seat 1–seat 4*.

Referring to Position 1 and 4 the differentiation of the word pairs and seat pairings are more distinctive compared to the Position 2 and 3. In general, the results of the

Table 6 Wilcoxon sign-rank test for Position 4

Position 4		Seat 1–seat 2	Seat 1–seat 3	Seat 1–seat 4	Seat 2–seat 3	Seat 2–seat 4	Seat 3–seat 4
Soft–hard	*Cushion*						
	Z	−1.255	−1.881	−1.588	−0.076	−2.931	−3.555
	P	0.209	0.060	0.112	0.940	**0.003**	**0.000**
	Backrest						
	Z	−1.807	−2.307	−1.900	−0.513	−3.006	−3.632
	P	0.071	**0.021**	0.057	0.608	**0.003**	**0.000**
Elastic–stiff	*Backrest*						
	Z	−0.867	−2.664	−1.380	−1.066	−1.794	−3.391
	P	0.386	**0.008**	0.168	0.286	0.073	**0.001**
Slippery–abrasive	*Cushion*						
	Z	−4.508	−2.029	−0.485	−3.947	−4.872	−2.895
	P	**0.000**	**0.042**	0.627	**0.000**	**0.000**	**0.004**
	Backrest						
	Z	−4.141	−2.504	−0.062	−2.617	−4.143	−2.617
	P	**0.000**	**0.012**	0.951	**0.008**	**0.000**	**0.008**

Wilcoxon signed-rank test show that the word pair *slippery–abrasive* is a differentiation factor independently from the position and the load. In contrast, the significance for the differentiation of the seats for the word pair *soft-hard* and *elastic-stiff* changes with the position.

3.1.3 Overall Rating

In Position 1 *seat 1* was rated as the best and *seat 4* as the worst seat. In Position 2 *seat 2* was rated as the best and *seat 4* as the worst seat. In Position 3 the best seat was *seat 3* and the worst one *seat 4*. Furthermore, in Position 4 *seat 1* was rated as the best and *seat 4* as the worst seat.

3.2 Objective Characterization of the Seats

3.2.1 Pressure Measurements

Table 7 shows the mean of all participants for each parameter: *average pressure*, *peak pressure*, and *contact area* for all four groups (*buttock Group, front Group, side Group* and *back Group*). The parameters with the highest values are made bold for each group. *Seat 1* (reference seat) has the highest *average pressure* and the highest

Table 7 Results of the pressure measurements. The highest values are bold for each group and parameter

		Seat 1		Seat 2		Seat 3		Seat 4	
		Mean	Std.	Mean	Std.	Mean	Std.	Mean	Std.
Buttock group	Average pressure [N/cm^2]	**0.50**	0.10	0.45	0.11	0.44	0.10	0.47	0.10
	Peak pressure [N/cm^2]	**1.20**	0.31	1.04	0.38	1.15	0.40	1.05	0.37
	Contact area [cm^2]	579	40	574	72	**643**	89	600	80
Front group	Average pressure [N/cm^2]	0.27	0.8	0.26	0.08	0.28	0.09	**0.32**	0.1
	Peak pressure [N/cm^2]	0.53	0.17	0.49	0.18	0.55	0.18	**0.63**	0.26
	Contact area [cm^2]	238	83	207	79	235	80	**264**	87
Side group	Average pressure [N/cm^2]	**0.28**	0.08	0.27	0.09	0.22	0.07	0.27	0.08
	Peak pressure [N/cm^2]	0.56	0.18	**0.59**	0.23	0.46	0.18	0.53	0.19
	Contact area [cm^2]	240	105	**240**	111	217	119	228	167
Back group	Average pressure [N/cm^2]	0.20	0.04	0.19	0.04	0.19	0.03	**0.21**	0.05
	Peak pressure [N/cm^2]	0.67	0.38	0.67	0.41	0.67	0.26	**0.69**	0.43
	Contact area [cm^2]	**530**	168	521	195	521	199	527	181

peak pressure in the area of the buttock (*buttock Group*). Especially the difference between the *peak pressure* of *seat 1* and *seat 4* is noticeable: even though *seat 1* has a seat suspension and *seat 4* a metal plate instead of the suspension, the *peak pressure* of *seat 1* is 0.15 N/cm^2 higher than for *seat 4*. For the *buttock Group seat 3* has the largest *area* in contact between person and seat. The measurement results for the front of the cushion (*front Group*), illustrate the highest *average pressure*, *peak pressure* and the largest *contact area* in *seat 4*. The lowest *average pressure*, *peak pressure*, and *contact area* has the Alcantara seat (*seat 2*). The values of *seat 1* and *seat 3* are close to the values of *seat 2*. Related to the bolster area of the cushion (*side Group*), the highest *average pressure* was found in *seat 1*, the highest *peak pressure* and contact *area* has the *seat 2*, and the lowest values for all three parameters has *seat 3*. The results regarding the backrest area (*back Group*) point out that the highest *average pressure* and the highest *peak pressure* is reached in *seat 4*. The largest *area* in contact between participant and the seat is found in *seat 1*.

In general, most of the measured differences between the four seats are small. The *peak pressure* reaches in the *buttock Group* the highest, in the *back Group* the second highest and in the *front Group* and *side Group* the lowest values. In addition, the *buttock Group* has the highest values for the *average pressure* and the *back Group* has the lowest values. The values of the *front Group* and *side Group* are in between those values.

3.2.2 Measurement Tool

Table 8 presents the results of the analysis of the four seats with the new developed measurement tool of Wegner et al. [19]. The results are divided into six blocks. Each block which contains the normalized values, compares the four seats through one appropriate measurement point. The detailed table without the normalized values is attached in the appendix. The maximum values are bold, and the minimum values are underlined.

The measurement results present that *seat 3* has the lowest pressure regarding the *first touch pressure* in cushion. As for the backrest, for most measurement points *seat 2* has the lowest *first touch pressure*. The *normalized linear pressure* (rise of pressure [N/cm²] per cm) appears in most measurement points for the backrest and the cushion the highest in seat 4, except for the lumbar area and the wings. In this measurement point *seat 1* shows the highest *normalized linear pressure* but the highest *linear indentation* at the same time. The *normalized maximum pressure* (pressure rises per cm until the maximum pressure is reached) is in *seat 4* the highest, except for the area of the ischial tuberosity. For this measurement point *seat 3* has the highest values. The lowest *normalized maximum pressure* has *seat 2*, except for the bolster in the backrest. *Seat 2* distributes the pressure (*pressure distribution*) the best for most measurement points. For the bolsters in the backrest and cushion *seat 3* distributes the pressure the most. The *normalized elongation while loading the seat* is for all measurement points for *seat 2* (Alcantara seat) the highest. The lowest *normalized elongation while loading the seat* has *seat 3*, except for the measurement point in the lumbar area and the backrest bolsters. For the lumbar *seat 1* and for the backrest bolster *seat 4* have the lowest *normalized elongation while loading the seat*. Concerning the *elongation due to the lateral movement seat 2* has the highest values in most cases. The highest *elongation due to the lateral movement* for the shoulder is evoked by *seat 4* and for the front of the cushion *seat 3* has the highest values. The *linear indentation* is for

Table 8 The Table illustrates the measurements results of the four seat in six measurement points. The highest values are highlighted bold numbers and the lowest values are highlighted with underlined numbers

	Max. pressure [N/cm² * 1/cm]	First touch [N/cm²]	Lin. pressure [N/cm² * 1/cm]	Pressure distribution [N/cm²]	Elongation loading [%/(N/cm²)]	Elongation move [%/(N/cm²)]	Max. indentation [mm]	Lin. indentation [mm]
(1) Shoulder								
Seat 1	3.5	0.60	0.9	0.70	1.07	0.99	31.9	16.1
Seat 2	2.1	**0.80**	1.0	**1.10**	**2.01**	1.05	**34.7**	**26.7**
Seat 3	3.1	0.40	0.7	0.80	0.77	0.83	**34.7**	14.8
Seat 4	**4.2**	0.60	**1.1**	0.80	1.35	**1.48**	31.4	13.2
(2) Lumbar								
Seat 1	2.9	**0.60**	**1.6**	0.50	0.06	**2.19**	33.2	**17.2**
Seat 2	1.2	0.40	0.8	**1.0**	**0.82**	**2.19**	**36**	15.7
Seat 3	1.8	0.50	0.7	0.70	0.59	1.92	35.7	14.4
Seat 4	**3.6**	0.50	0.9	0.50	0.14	1.85	30.1	12.3
(3) Bolster backrest								
Seat 1	7.6	**0.80**	1.7	0.60	0.49	2.69	23.4	12
Seat 2	6.0	0.40	0.9	0.80	**0.55**	**2.81**	26.1	12.7
Seat 3	5.5	0.60	1.1	**0.90**	0.47	2.50	24.7	**14.2**
Seat 4	**7.7**	0.60	**2.2**	0.80	0.24	2.57	24.3	13.0
(4) Ischial tuberosity								
Seat 1	3.3	0.70	1.7	0.50	0.42	1.86	30.1	**22.4**
Seat 2	1.1	0.60	0.8	**0.90**	**1.32**	**2.59**	31.5	19.6
Seat 3	**4.1**	0.60	1.0	0.40	0.36	1.78	30.3	12.7

(continued)

Table 8 (continued)

	Max. pressure [N/cm² * 1/cm]	First touch [N/cm²]	Lin. pressure [N/cm² * 1/cm]	Pressure distribution [N/cm²]	Elongation loading [%/(N/cm²)]	Elongation move [%/(N/cm²)]	Max. indentation [mm]	Lin. indentation [mm]
Seat 4	3.8	**0.80**	**1.9**	0.40	0.53	1.58	30.3	18.9
(5) Front of the cushion								
Seat 1	4.1	0.80	**2.7**	0.30	0.30	1.77	28.5	**22.8**
Seat 2	2.1	0.80	1.2	**0.80**	**0.60**	2.23	29.7	16.9
Seat 3	4.8	0.50	1.0	0.40	0.08	**2.25**	28.9	14.3
Seat 4	**8.9**	**0.90**	**2.7**	0.20	0.29	0.55	29	16.2
(6) Bolster cushion								
Seat 1	3.3	0.50	**0.90**	0.60	0.89	3.66	30.4	**16.2**
Seat 2	2.5	**0.60**	0.80	**0.90**	**1.05**	**4.58**	28.8	19
Seat 3	3.7	0.20	0.80	**0.90**	0.59	3.73	28.6	12.5
Seat 4	**4.2**	0.40	0.70	0.80	0.88	4.55	24.3	15.4

Table 9 Overview of the static and dynamic friction coefficient for various material parings

	μ_{static}	$\mu_{dynamic}$
Leather–silicon	–	1.38
Alcantara–silicon	–	1.30
Leather–jeans	0.35	0.34
Alcantara–jeans	1.03	0.70

seat 3 the lowest and for *seat 1* the highest regarding the cushion. The lowest *linear indentation* mostly has seat 4 in reference to the backrest.

In summary, the results show that *seat 4* can be identified as hardest regarding the pressure measurements with the new tool and *seat 2* and *3* the softest. *Seat 1* is in between. While loading the seat, *seat 2* shows the most elongation. *Seat 3* has the least elongation recorded by the stamp sensors (I–IV, Fig. 4) or rather elongate the human skin. *Seat 3* shows also the least linear characteristics (*linear indentation* is the lowest) and *seat 1* has the most. Considering the backrest *seat 4* has the lowest linear properties.

3.2.3 Friction Measurement

Table 9 presents the results of the friction tests. The *leather–silicon* and *Alcantara–silicon* combination showed no static friction even with forces over 100 N (the force used in all test) the combination skips immediately to sliding. The dynamic μ is for the *leather–silicon* combination a bit higher than for *Alcantara–silicon*. For the jeans combinations with leather and Alcantara a static μ could be detected. The μ_{static} is for a leather cover three times lower than for Alcantara, the $\mu_{dynamic}$ is nearly two times lower. The friction coefficient for *leather–jeans* is nearly the same for static and dynamic setups.

3.2.4 The Influence of the Friction Coefficient on the Measurement Data

The *elongation while loading the seat* and the *elongation due to the lateral movement* recorded by the stamp are based on the friction coefficient including silicon (*leather–silicon* and *Alcantara–silicon*). To include also clothing materials like jeans, which are in direct contact with the seat surface, theses parameters were normalized based on the dynamic friction coefficient of silicon (see Sect. 2.4.2 and Table 8) and afterwards multiplied with the dynamic friction coefficient of the jeans pairings. Table 10 presents the results exemplary for the cushion. The highest values are bold and the lowest are underlined.

The calculated parameter *elongation while loading the seat* and *elongation due to the lateral movement* for the jeans pairings (*leather–jeans* and *Alcantara–jeans*) are for each measurement point the highest in *seat 2*. The *elongation while loading*

Table 10 Results of the parameters elongation while loading and elongation due to the lateral movement including the interaction with a jeans material. The highest values are bold and lowest underlined

	Seat 1	Seat 2	Seat 3	Seat 4	Seat 1	Seat 2	Seat 3	Seat 4	Seat 1	Seat 2	Seat 3	Seat 4
	(4) Ischial tuberosity				(5) Front of the cushion				(6) Bolster cushion			
Elongation loading [%/(N/cm^2)]	0.14	**0.93**	0.12	0.18	0.10	**0.42**	0.03	0.10	0.30	**0.75**	0.20	0.20
Elongation move [%/(N/cm^2)]	0.63	**1.81**	0.61	0.54	0.60	**1.56**	0.76	0.19	1.24	**3.21**	1.27	1.06

is for each measurement point of the cushion in *seat 3* the least. For the *elongation due to the lateral movement* the lowest values are found for all measurement points in *seat 4*.

4 Discussion

4.1 *Discomfort of the Seats*

The discomfort ratings have shown that the seat components (foam, seat cover, seat suspension) of the reference seat (*seat 1*) are more balanced than the manipulated seats (*seat 2, seat 3, seat 4*). In particular, in the sensitive shoulder area [18] the participants perceived discomfort on the outer edge of the manipulated seats. The reason might be that a disharmony is perceived, meaning that particular parts of the seat do not match with other parts of the seat while sitting. Neither the pressure measurement nor the results of the measurement tool have data that clearly explain the discomfort in these parts. The pressure distribution of the participants, who stated discomfort in those areas, had no pressure peaks or points. The measurement tool did not measure remarkable characteristics in this particular area; therefore, exact predictions and explanation are hard to make.

Seat 2, the Alcantara seat, has noticeable discomfort ratings in the rear bolster of the cushion. The implemented shear force through the higher friction coefficient (*leather–jeans* vs. *Alcantara–jeans*) might cause an additional force which results in a discomfort feeling. This is in line with Chow and Odell [2] who linked the pressure perception to the shear force perception. Furthermore, the measurement tool of Wegner et al. [19] confirms this perception. The measurement results in the bolster show a low pressure, but large elongations and therefore additional tensile strain might be felt, which could also evoke the shear force [8]. The explanation why only a few participants rated this as discomfort could be that some of these participants are shear sensitive or because the hips of the participants were wider. Another reason could be that the combinations of pressure and shear evokes a discomfort feeling [2].

Seat 3 has a noticeable discomfort in the bolsters of the backrest. The loose cover tension might evoke that some of the participants sink more into the backrest of the seat and thus feel the plastic plates of the side adjustments because of the higher compression of the foam. The pressure measurement of the participants does not explain the discomfort ratings, because the values of *seat 3* are not significantly different to the pressure values of *seat 1, seat 2* and *seat 4*. The pressure mat itself could influence the measurement by its thickness negatively and therefore might not record the pressure distribution in the right way. The results of the measurement tool illustrate that the bolster of *seat 3* has nearly the same pressure characteristics as the other seats. In combination with a softer lumbar area (higher tendency to sink into the seat) there is more contact with the bolsters and these are more compressed and therefore might be perceived more uncomfortable.

Seat 4 has a noticeable discomfort in the lumbar area. Replacing the seat suspension with a metal plate could influence this lumbar support experience, affecting the sitting posture not supporting the natural S-shape of the spine. This might be the reason why some participants perceived a discomfort in this region of the body. For the lumbar area the pressure measurements correspond to the experience. The *average pressure* and the *peak pressure* are both the highest in this area. The results of the measurement tool show that for higher loading the pressure for *seat 4* rises the most and has the highest value of all four seats.

Overall, with the help of the measurement tool it is possible to explain the discomfort rating better than the results of the pressure measurements alone. The correlation of the pressure mat measurements and discomfort is useful for only some parts (lumbar area) of the seat: in most cases the recordings of the pressure mat are not useful for building a correlation because the pressure mat does not record the influence of the surface or even the tension of the cover leading to elongation and shear force. In contrast, the measurement tool records the influences of the surface and the cover tension but is only capable to measure in discrete points. In future it would be good to study the connection of the measurement points to what is happening in the human seat interface at that point and connected to pressure mat measurements for more precise statements.

4.2   Characterization

Based on the word pair ratings, the results have shown that the Alcantara seat (*seat 2*) is characterized abrasive so that the abrasive surface differentiates *seat 2* in each position from the other seats. The differentiation of the surface might be independent from the position or the loading due to the significant results of the surface differentiation in each position. The pressure measurement does not record this perceived difference of the surface but the results of the measurement tool illustrate the difference presenting the highest *elongation while loading the seat* in each measurement point for the abrasive *seat 2*. In addition, the parameter *normalized elongation due to the lateral movement* is for the material pairing *silicon–Alcantara* for nearly every measurement point the highest. That is in line with the study of Goossens [7] who stated that a LiquiCell cushion material evokes less shear stress (internal shear stress) than a foam cushion. Including the friction coefficient of all jeans-pairings demonstrate the high impact of the external applied shear force provoked by high friction coefficients. Therefore, the adapted parameter of Table 10 *elongation due to the lateral movement* including the friction coefficients of the jeans pairings (*Alcantara–jeans* and *leather–jeans*) have in each measurement point the highest shear force in *seat 2*. These measurement results are in line with the perceived differences of the participants.

The results of the word pair rating for the hardness and elasticity show that the differentiation of both parameters depends on the position. For Position 1 and Position

4 the participants differentiate the hardness and elasticity of the seats most significantly. In both positions the cushion angle is high ($15°$–$18°$). Therefore, the sensitive
area of the body (front of the cushion) is in contact with the seat [18] and might be
the reason for the differentiation. In contrast, Position 2 and 3 with a cushion angle
of $3°$ and hence less sensitive contact area in the front of the cushion the participants
notice less differences regarding the hardness and elasticity of the seats. Additional,
in Position 1 the backrest is more upright than in Position 4 (but both high cushion
angles). Therefore, the differentiation of the cushion is probably more related to the
area being in contact than to the load. In Position 2 and 3 the participants were not
able to differentiate the elasticity neither for the backrest nor the cushion. *Seat 4*
characterized as the hardest seat differs in Position 2 regarding the cushion hardness
from the other seat, but in Position 3 with the same cushion angle but a more horizontal backrest angle (less load on cushion) the hardness of the cushion cannot be
differentiated anymore. Furthermore, the more horizontal angle in Position 3 than
in Position 2 evokes a higher contact area with the sensitive shoulder area (same
cushion angle). The results of the word pair rating suggest that in Position 3 the
hardness of the backrest can be differentiated, whereas in Position 2 it cannot be
differentiated. Therefore, it might be concluded that also the sensitive areas of the
backrest evoke a better differentiation of the seats. Position 1 and 4 are the most
significant positions regarding the hardness and the elasticity differentiation. *Seat 3*
with the loose cover tension and therefore with the best foam properties was rated
as the most elastic seat. Unfortunately, the pressure mat measurements do not offer
a connection to the seat elasticity but the results of the measurement tool of Wegner
et al. [19] demonstrate that the parameter *normalized elongation while loading the
seat* is in almost every measurement point (except lumbar and backrest bolster) the
lowest. Therefore, *seat 3* might not stress the skin as much as in the other seats.
On the one hand, the low cover tension provokes the best foam properties and thus,
the best spring/damper properties. On the other hand, it causes a high interaction
between the seat suspension and the foam. This fact is illustrated by the results of
the measurement tool in the measurement point of the ischial tuberosity. The results
present the highest maximum pressure in *seat 3* hence to a high relative movement
between the seat suspension and the foam. The foam is pressing through the suspension spring. For all other measurement points the results of stamp measurements
show that seats 2 and 3 both have the lowest pressure attributes or rather the best
pressure distributions. On the contrary, *seat 4*, characterized as the hardest seat, has
the highest maximum pressure in each measurement point and an unequal pressure
distribution. The results of the measurement tool are in line with the results of the
word pair ratings. Unfortunately, the results of the pressure mat measurements do
not correlate with the results of the word pair ratings in most cases.

However, for further studies the different loadings and the connection of the
different measurement points of the measurement tool should be taken more into
account. The study has shown that the position, the contact area, and the sensitivities
of the human body influence the ratings and the characterizations of a seat. This
should additionally be included into the measurement procedure of the measurement

tool. Moreover, it is pointed out that the optimum position for an occupant in one specific seat is not necessarily the optimum position in another seat with different cover and seat suspension properties.

5 Conclusion

The study has shown that seats with the same contour and foam properties and differ in cover (surface and cover tension) and seat suspension are perceived different. The seat layout has a huge impact on the seat-human interaction and therefore influences the parameters for the seat characterization. Moreover, the positions evoke various significances for the differentiation due to different sensitivity areas in contact with the seat. The results of the objective measurement tool from Wegner et al. [19] could be used to explain the rated characteristics of the seats. The correlations between the discomfort ratings and the stamp measurements could be improved by including the mutual interdependencies of the measurement points. Unfortunately, in most cases the pressure mat measurements neither correlate with the discomfort rating nor with the characterizations of the seats. In order to receive a more precise characterization as well as a more precise discomfort rating the results of the measurement points (measurement tool Wegner et al. [19]) and the interdependencies of the measured parameter have to be correlated and evaluated in further studies with various participants and seats.

Acknowledgments We are thankful to our TU Delft colleagues and Rudolf Rackl, Renato Martic and Mario Buljan who provided expertise that greatly assisted the research. We are also grateful to Erich Zerhoch and Johann Sax for assistance organizing the measuring instruments and for assistance building up the setup.

Appendix

Descriptive Results of the word pair ratings for Position 1–Position 4

Position 1	Seat 1		Seat 2		Seat 3		Seat 4	
	Mean	Std. dev.	Mean	Std. dev.	Mean	Std. dev.	Mean	Std. dev.
Cushion								
Soft–hard	3.9	1.4	3.9	1.4	3.6	1.4	4.5	1.4
Elastic–stiff	4.3	1.3	4.4	1.2	3.7	1.5	4.4	1.5
Slippery–abrasive	3.7	1.4	4.4	1.4	3.6	1.7	3.2	1.5
Backrest								
Soft–hard	4.2	1.4	4.0	1.4	3.7	1.3	4.8	1.4

(continued)

(continued)

Position 1	Seat 1		Seat 2		Seat 3		Seat 4	
	Mean	Std. dev.	Mean	Std. dev.	Mean	Std. dev.	Mean	Std. dev.
Elastic–stiff	4.6	1.3	4.7	1.4	3.9	1.5	4.9	1.5
Slippery–abrasive	3.7	1.5	5.5	1.4	4.2	1.5	4.0	1.5

Position 2	Seat 1		Seat 2		Seat 3		Seat 4	
	Mean	Std. dev.	Mean	Std. dev.	Mean	Std. dev.	Mean	Std. dev.
Cushion								
Soft–hard	3.5	1.3	3.3	1.4	3.6	1.4	4.0	1.4
Elastic–stiff	3.7	1.2	3.9	1.4	3.8	1.3	4.1	1.4
Slippery–abrasive	3.8	1.6	5.7	1.1	4.1	1.4	3.5	1.4
Backrest								
Soft–hard	4.1	1.2	4.0	1.4	4.0	1.2	4.5	1.4
Elastic–stiff	3.7	1.2	3.9	1.3	3.8	1.3	4.1	1.4
Slippery–abrasive	3.9	1.3	5.6	1.0	4.3	1.3	3.8	1.3

Position 3	Seat 1		Seat 2		Seat 3		Seat 4	
	Mean	Std. dev.	Mean	Std. dev.	Mean	Std. dev.	Mean	Std. dev.
Cushion								
Soft–hard	3.6	1.4	3.5	1.3	3.7	1.3	4.0	1.5
Elastic–stiff	3.7	1.4	3.7	1.4	3.9	1.2	3.9	1.4
Slippery–abrasive	3.4	1.4	5.6	0.9	3.8	1.3	3.5	1.3
Backrest								
Soft–hard	4.6	1.2	4.5	1.3	4.2	1.3	4.9	1.3
Elastic–stiff	4.4	1.3	4.7	1.2	4.4	1.2	4.7	1.3
Slippery-abrasive	4.0	1.4	5.6	0.9	4.1	1.4	4.1	1.3

Position 4	Seat 1		Seat 2		Seat 3		Seat 4	
	Mean	Std. dev.	Mean	Std. dev.	Mean	Std. dev.	Mean	Std. dev.
Cushion								
Soft–hard	4.0	1.5	3.7	1.3	3.6	1.5	4.4	1.3
Elastic–stiff	4.1	1.4	4.1	1.2	3.9	1.4	4.5	1.5
Slippery–abrasive	3.6	1.6	5.5	1.2	4.11	1.3	3.4	1.5
Backrest								
Soft–hard	4.7	1.5	4.1	1.4	4.2	1.2	5.2	1.4

| Elastic–stiff | 5.0 | 1.1 | 4.8 | 1.3 | 4.5 | 1.2 | 5.3 | 1.3 |
| Slippery–abrasive | 4.0 | 1.4 | 5.3 | 1.1 | 4.6 | 1.3 | 3.9 | 1.3 |

Overview of the non-normalized measurement results of the four seat in six different measurement positions of the seas.

	Max. pressure [N/cm^2]	First touch [N/cm^2]	Lin. pressure [N/cm^2]	Pressure distribution [N/cm^2]	Elongation loading [%]	Elongation move [%]	Max. indentation [mm]	Lin. indentation [mm]
(1) Shoulder								
Seat 1	11.10	0.60	1.50	0.70	16.40	15.20	31.90	16.10
Seat 2	7.20	0.80	2.80	1.10	18.80	9.80	34.70	26.70
Seat 3	10.60	0.40	1.10	0.80	11.20	12.10	34.70	14.80
Seat 4	13.20	0.60	1.40	0.80	24.60	27.00	31.40	13.20
(2) Lumbar								
Seat 1	9.60	0.60	2.80	0.50	0.80	29.00	33.20	17.20
Seat 2	4.40	0.40	1.20	1.00	4.70	12.50	36.00	15.70
Seat 3	6.40	0.50	1.00	0.70	5.20	17.00	35.70	14.40
Seat 4	10.80	0.50	1.10	0.50	2.10	27.50	30.10	12.30
(3) Bolster backrest								
Seat 1	17.70	0.80	2.00	0.60	12.00	65.80	23.40	12.00
Seat 2	15.70	0.40	1.20	0.80	11.30	57.30	26.10	12.70
Seat 3	13.70	0.60	1.60	0.90	8.80	47.30	24.70	14.20
Seat 4	18.80	0.60	2.90	0.80	6.10	66.60	24.30	13.00
(4) Ischial tuberosity								
Seat 1	10.00	0.70	3.80	0.50	5.80	25.70	30.10	22.40
Seat 2	3.60	0.60	1.60	0.90	6.20	12.10	31.50	19.60
Seat 3	12.40	0.60	1.30	0.40	6.10	30.50	30.30	12.70
Seat 4	11.50	0.80	3.50	0.40	8.40	25.10	30.30	18.90

(continued)

(continued)

	Max. pressure [N/cm²]	First touch [N/cm²]	Lin. pressure [N/cm²]	Pressure distribution [N/cm²]	Elongation loading [%]	Elongation move [%]	Max. indentation [mm]	Lin. indentation [mm]
(5) Front of the cushion								
Seat 1	11.60	0.80	6.20	0.30	4.80	28.30	28.50	22.80
Seat 2	6.30	0.80	2.00	0.80	4.90	18.30	29.70	16.90
Seat 3	14.00	0.50	1.50	0.40	1.50	43.40	28.90	14.30
Seat 4	25.70	0.90	4.40	0.20	10.40	19.50	29.00	16.20
(6) Bolster cushion								
Seat 1	10.10	0.50	1.40	0.60	12.40	51.00	30.40	16.20
Seat 2	7.30	0.60	1.60	0.90	10.00	43.50	28.80	19.00
Seat 3	10.70	0.20	1.00	0.90	8.70	55.10	28.60	12.50
Seat 4	10.10	0.40	1.10	0.80	8.40	43.30	24.30	15.40

References

1. Biedermann H, Guttmann G (1984) Funktionelle Pathologie und Klinik der Wirbelsäule, Band 1, Teil 2, Elsevier, Munich
2. Chow WW, Odell EI (1978) Deformations and stresses in soft body tissues of a sitting person
3. DIN 53579 (2005) Testing of flexible cellular materials—Indentation test on finished parts. DIN-Normenausschuss Materialprüfung (NMP)
4. DIN EN ISO 3386-1 (2009) Polymeric materials, cellular flexible—Determination of stress-strain characteristics in compression—Part 1: low-density materials. DIN-Normenausschuss Materialprüfung (NMP)
5. Goossens RHM, Snijders CJ (1995) Design cirteria for the reduction of shear forces in beds and seats. J Biomech 28(2):225–230
6. Grujicic M, Pandurangan B, Arakere G, Bell WC, He T, Xie X (2009) Seat-cushion and soft-tissue material modeling and a finite element investigation of the seating comfort for passenger-vehicle occupants. Mater Des 30(10):4273–4285. https://doi.org/10.1016/j.matdes. 2009.04.028
7. Goossens RHM (2001) Shear stress measured on three different cushioning materials. Delft University of Technology
8. Goossens RHM, Teeuw R, Snijders CJ (2000) Decubitus risk: is shear more important than pressure? In: Proceedings of the IEA 2000 IHFES 2000 Congress
9. Hartung J (2006) Objektivierung des statischen Sitzkomforts auf Fahrzeugsitzen durch die Kontaktkräfte zwischen Mensch und Sitz. Dissertation, Technischen Universität München

10. Harrison DD, Harrison SO, Croft AC, Harrison DE, Troyanovich SJ (2000) Sitting Biome-chanics Part II: optimal car driver's seat and optimal driver's spinal model. J Manipulative Physiol Ther 23(1):37–47
11. Hiemstra-van Mastrigt S (2015) Comfortable passenger seats: recommendations for design and research. PhD dissertation, Delft University of Technology
12. Kilincsoy Ü (2015) Digitalization of posture-based seat design. PhD dissertation, Delft University of Technology
13. Mansfield N, Sammonds G, Nguyen L (2015) Driver discomfort in vehicle seats- Effect of changing road conditions and seat foam compositions. Appl Ergonom 50:153–159
14. McCrady SK, Levine JA (2009) Sedentariness at work: how much do we really sit? Obesity 17(11):2103–2105
15. Molenbroek JFM, Albin TJ, Vink P (2017) Thirty years of anthropometric changes relevant to the width and depth of transportation seating spaces, present and future. Appl Ergonom 65:130–138. https://doi.org/10.1016/j.apergo.2017.06.003
16. Sammonds GM, Fray M, Mansfield NJ (2017) Effect of long term driving on driver discomfort and its relationship with seat fidgets and movements (SFMs). Appl Ergonom 58:119–127
17. Schmidt RF, Thews G (1980) Physiologie des Menschen. Springer Verlag, Berlin
18. Vink P, Lips D (2017) Sensitivity of the human back and buttocks: The missing link in comfort seat design. Appl Ergonom 58:287–292
19. Wegner M, Martic M, Franz M, Vink P (2017) A new approach for measuring the sea comfort. In: 11th international comfort congress, Salerno
20. Wegner M, Anjani S, Li W, Vink P (2019) How does the seat cover influence the seat comfort evaluation? In: Bagnara S, Tartaglia R, Albolino S, Alexander T, Fujita Y (eds) Proceedings of the 20th congress of the international ergonomics association (IEA 2018). IEA 2018. Advances in intelligent systems and computing, vol 824. Springer, Cham
21. Zenk R, Mergl C, Hartung J, Sabbah O, Bubb H (2006) Objectifying the comfort of car seats. SAE International
22. Zuo H, Jones M, Hope T (2004) A matrix of material representation. In: Proceedings of future ground, design research society international conference. Melborne

Vibro-Acoustic Metamaterials for Improved Interior NVH Performance in Vehicles

Lucas Van Belle, Luca Sangiuliano, Noé Geraldo Rocha de Melo Filho,
Matias Clasing Villanueva, Régis Boukadia, Sepide Ahsani, Felipe Alves Pires,
Ze Zhang, Claus Claeys, Elke Deckers, Bert Pluymers, and Wim Desmet

Abstract Due to environmental and economic requirements, lightweight design
is increasingly used in automotive applications. However, the use of lightweight
design typically comes at the cost of impaired NVH performance. Classic solutions
to improve the NVH performance usually rely on adding mass or volume, conflicting with lightweight design. In the search for innovative lightweight and compact
solutions for noise and vibration reduction, vibro-acoustic locally resonant metamaterials have recently emerged. Through the introduction of local resonances in
a flexible host structure on a sub-wavelength scale, frequency ranges without free
wave propagation can be created, referred to as stop bands. These stop bands enable
achieving targeted frequency ranges of strong noise and vibration attenuation, while
the sub-wavelength nature of locally resonant metamaterials enables lighter and thinner vibro-acoustic solutions, which are also able to target the hard-to-address low-
frequency range. Since their emergence, the potential of locally resonant metamaterials has been demonstrated and their application to automotive NVH problems has
attracted increasing attention. This chapter gives an overview of the vibro-acoustic
locally resonant metamaterial research, discusses the potential of these metamaterials
for automotive applications and presents a locally resonant metamaterial application
for interior noise reduction in a targeted frequency range in a real vehicle.

L. Van Belle (✉) · L. Sangiuliano · N. G. Rocha de Melo Filho · M. Clasing Villanueva ·
R. Boukadia · S. Ahsani · F. Alves Pires · Z. Zhang · C. Claeys · B. Pluymers · W. Desmet
DMMS Lab, Flanders Make, KU Leuven, Department of Mechanical Engineering,
Celestijnenlaan 300, Heverlee, Belgium
e-mail: lucas.vanbelle@kuleuven.be

Division LMSD, KU Leuven, Department of Mechanical Engineering, Celestijnenlaan 300,
Heverlee, Belgium

E. Deckers
DMMS Lab, Flanders Make, KU Leuven, Campus Diepenbeek, Department of Mechanical
Engineering, Wetenschapspark 27, Diepenbeek, Belgium

Division LMSD, KU Leuven, Campus Diepenbeek, Department of Mechanical Engineering,
Wetenschapspark 27, Diepenbeek, Belgium

A. Fuchs and B. Brandstätter (eds.), *Future Interior Concepts*,
Automotive Engineering : Simulation and Validation Methods,
https://doi.org/10.1007/978-3-030-51044-2_2

1   Introduction

Ever tightening environmental legislation and increasing economic requirements
have given rise to the introduction of novel, cleaner propulsion systems and to the
use of lightweight design in the transportation industry, in order to improve fuel
economy and to reduce emissions [1]. To enhance structural integrity, reliability,
drivability and passenger safety, lightweight structures are typically also designed
for higher strength and stiffness. However, their resulting increased stiffness-to-mass
ratio leads to an impaired noise, vibration and harshness (NVH) performance.

The intrinsically reduced NVH performance of lightweight structures conflicts
with the increasingly stringent noise exposure regulations and customer expectations
regarding quality and comfort [2]. The latter is also gaining importance in electric
vehicles, since replacing the internal combustion engine causes associated beneficial
masking effects to disappear, which causes squeak and rattle, wind and tire noise
to become more noticeable, while the electric drivetrain introduces different noise
spectra [3]. To restore the NVH performance, add-on treatments are often resorted
to. Classic passive noise control treatments typically rely on adding mass to increase
the sound transmission loss by exploiting the acoustic mass-law or adding volume
to increase the sound absorption. The latter also becomes prohibitive in the hard-
to-address low-frequency range since an absorptive layer thickness of at least one
quarter wavelength is typically required to adequately dissipate the sound energy.
While active noise control can in some cases perform better at low frequencies, it also
requires mass and volume additions, as well as an energy supply. The conventional
approaches typically lead to heavy and bulky solutions, conflicting with the trend
towards lightweight design, or are inadequate for low frequencies.

To face the shortcomings of classic noise control solutions, novel material concepts are pursued that can simultaneously satisfy ecologic and economic requirements as well as noise exposure regulations and customer expectations. A vast
amount of research has emerged in innovative solutions with favourable vibro-
acoustic behavior to try and comply with these conflicting requirements. Ever thinner
and lighter solutions are sought, which are also suitable to target the hard-to-address
low-frequency range. Among these, locally resonant metamaterials (LRMs) have
come to the fore around the beginning of this century. In general, LRMs are artificial
structures that are designed to have unusual properties which go beyond naturally
occurring or conventional materials [4, 5]. The LRM research originated in the field
of electromagnetics [6] and later extended to acoustics and elastodynamics [7, 8].
Extraordinary functionalities such as negative refraction, cloaking, focusing, trap-
ping and waveguiding emerged in different fields [5, 7, 9]. For noise and vibration
control engineering, one of the appealing properties of LRMs is their ability to create
stop bands. These are frequency ranges in which no free wave propagation is allowed
[10], which can hence enable strong attenuation of vibrations and noise.

Stop bands in LRMs arise from the introduction of local resonances in a flexible
host structure on a sub-wavelength scale. The sub-wavelength nature of LRMs can
enable lighter and thinner vibro-acoustic solutions, also able to target the hard-to-

address lower frequencies. Since their emergence, the potential of LRMs for noise control in specific frequency ranges has been widely demonstrated. In recent years, their application to automotive NVH problems has attracted increasing attention and promising first demonstrations have been presented. In this chapter, the emergence of LRMs for vibro-acoustics is discussed and their potential for interior vehicle noise reduction in a targeted frequency range is demonstrated by means of an application.

The chapter is organized as follows. Section 2 gives an overview of the state of the art of the vibro-acoustic LRM research. Section 3 discusses recent examples of their potential value in an automotive context. Section 4 presents the design and experimental validation of an LRM concept applied to the rear shock towers of a vehicle in order to reduce interior noise. Eventually, Sect. 5 summarizes the main conclusions of this chapter.

2 The Emergence of Vibro-Acoustic LRMs: An Overview

The notion of LRMs originated in the field of electromagnetics during last century [6]. Inspired by the quantum mechanical band theory, describing the existence of energy bands in which electrons in solids can reside, separated by forbidden band gaps, academic curiosity has driven the search for band gaps in other fields in order to manipulate wave propagation. Photonic crystals were proposed, which are periodic structures with band gaps for electromagnetic wave propagation [11, 12]. Band gaps, or stop bands, are frequency ranges in which no free wave propagation is allowed [10]. The analogies between the electromagnetic, acoustic and elastic wave equations have led to the emergence of phononic crystals with stop bands for acoustic and for elastic wave propagation [13, 14]. The stop bands in photonic and phononic crystals arise from the periodic arrangements of the constituents with high impedance contrast. This leads to Bragg scattering for wavelengths comparable to the characteristic periodicity length scale, causing destructive interference between incoming and scattered waves [15]. The length scale dependency of these interference-based stop bands, however, limits their suitability for low frequencies, for which prohibitively large and bulky structures would again be required [5, 9].

This length scale dependency was resolved by the advent of LRMs. Already in the late 1960s the concept of LRMs was introduced in the field of electromagnetics by Veselago [6], to be brought into practice by Pendry et al. [16] only in the late 1990s. By introducing local resonances, realized by split-ring metallic resonators, in a periodic array, a stop band was obtained around the tuned resonance frequency of the resonators, far below the Bragg interference limit. Inspired by the advances in electromagnetics, LRMs for acoustic and for elastic wave manipulation emerged around the turn of the century [8, 17]. The stop bands in LRMs do not require periodicity, but arise from the sub-wavelength addition of local resonators to a hosting medium and result from a Fano-type interference between incoming waves and out-of-phase re-radiated waves by the local resonators [18].

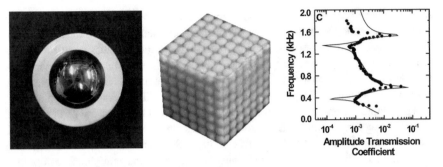

Fig. 1 Rubber-coated lead sphere resonator (left) of the *locally resonant sonic crystal* introduced by Liu et al. [8] (center) and the measured (dots) and predicted (line) acoustic amplitude transmission coefficient through the sample (right) showing zones of strong transmission reduction [5]

The potential of LRMs for vibro-acoustics was first demonstrated by Liu et al. [8]. Based on an elastic phononic crystal design, local resonances were introduced by inserting rubber coated lead spheres with a cubic periodic arrangement in an epoxy matrix (Fig. 1). This *locally resonant sonic crystal* exhibited resonance-based stop bands for elastic wave propagation around the resonance frequencies of the resonators, well below the Bragg interference limit of the lattice. Upon acoustic excitation, strong attenuation through the LRM sample was obtained in the audible frequency range inside the stop bands, outperforming the acoustic mass-law. This was attributed to a negative dynamic effective mass, resulting in exponential attenuation of the elastic wave propagation in the stop band [19]. Given the sub-wavelength arrangement of the resonant inclusions, the resonance-based stop band was shown not to depend on periodicity. Liu et al. [8] also measured the acoustic transmission through an LRM sample of a single layer with randomly, sub-wavelength arranged resonant inclusions, in an impedance tube. Similar acoustic insulation improvements were measured inside the stop bands, demonstrating the potential of vibro-acoustic LRMs relying on an LRM structure with resonance-based stop band behavior for elastic wave propagation to improve the acoustic insulation in a targeted frequency range. This opened the door towards thinner and lighter solutions for vibro-acoustic problems in specific frequency bands, also able to target the low-frequency range.

Driven by the need for lighter and thinner vibro-acoustic solutions, an interest grew in designing elastic LRM partitions for improved acoustic insulation. Inspired by the design of Liu et al. [8], the sound transmission loss (STL) of an elastic LRM slab composed of a single layer of silicon rubber-coated steel balls in a polyester matrix was investigated by Calius et al. [20]. Ho et al. [21] measured STL improvements of a similar elastic LRM partition, in which resonant cells composed of rubber-coated steel balls were assembled in a rigid plastic grid host structure, consisting of one or multiple stacked layers. In view of practical noise and vibration engineering solutions, the research in resonance-based stop bands in LRM plates further expanded, with a growing interest in targeting the acoustically relevant out-of-plane vibrations. The stop band behavior of LRM plates with 2D periodic arrangements of rubber

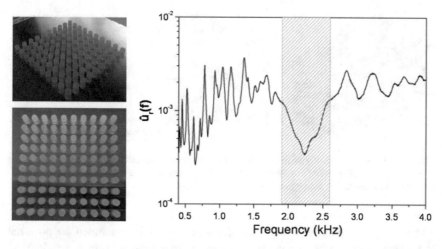

Fig. 2 LRM plate with a 2D periodic array of silicon rubber stub resonators (left) and measured average out-of-plane displacement response for a structural excitation (right) showing a frequency range of strong reduction [28]

disc inclusions [22], rubber-coated steel disc inclusions [23] and hemmed discs [24] was investigated. Aside from LRM structures with resonant inclusions, the addition of resonators onto plate host structures was investigated. Xiao et al. [25] and Claeys et al. [26] analyzed the sub-wavelength, periodic addition of idealized mass-spring resonators to plates in order to create bending wave stop bands. Wu et al. [27] proposed an LRM plate consisting of a 2D periodic array of cylindrical aluminium stub resonators on an aluminium plate to obtain bending wave stop bands. Oudich et al. [28] considered less stiff, rubber stub resonators, in view of obtaining vibration attenuation at lower frequencies (Fig. 2). This was further elaborated by Assouar et al. [29], who added masses on top of the rubber stubs to lower and broaden the stop band by increasing the resonant mass. Xiao et al. [30] proposed an LRM plate consisting of a 2D periodic array of beam-shaped resonators attached to a thin plate to obtain a resonance-based bending wave stop band. Van Belle et al. [31] added a 2D periodic array of laser-cut PMMA beam-shaped resonators to an aluminum plate and measured the impact of damping on the stop band behavior. Mainly damping in the resonators was found to affect the stop band, reducing peak attenuation, but increasing vibration reduction in a broadening frequency range around the stop band. Going towards lightweight design integration, Claeys et al. [32] proposed a 3D printed LRM structure consisting of a honeycomb core with three-legged resonators added in the hollow core cavities, to create a low-frequency bending wave stop band.

Also the stop band behavior in LRM beams and structural waveguides was widely investigated [26, 33, 34]. Claeys et al. [35] investigated the creation of multiple structural stop bands in square cross-section ducts by combining differently tuned resonators, to reduce the vibration transmission in a broadened frequency band along

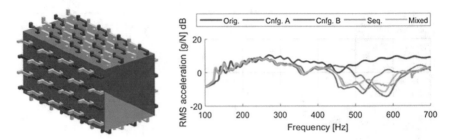

Fig. 3 Structural waveguide with periodic beam-like resonators (left) and measured vibration transmission along the waveguide for different resonator configurations and resonator tuning (right) for a structural excitation at one end [35]

a known transmission path (Fig. 3). Nateghi et al. [36] demonstrated the potential of LRMs to reduce vibration transmission in and sound radiation from a cylindrical waveguide by adding laser-cut beam-shaped resonators on a sub-wavelength scale.

The exploitation of resonance-based stop bands for out-of-plane bending waves in LRM panels to reduce acoustic radiation and transmission was investigated. Claeys et al. [37] reported strong acoustic radiation reduction inside the bending wave stop band of LRM plates with periodic mass-spring resonators for a structural excitation, albeit at the cost of a potential additional coincidence zone at the end of the stop band. The STL of LRM plates with periodic mass-spring resonators was extensively analyzed [38–40]. Inside the stop band, a strong mass-law outperforming STL peak can be obtained. This was found to correspond to a strong increase in the effective dynamic mass of the LRM plate near the resonance frequency of the resonators [40, 41]. However, an STL dip typically follows around the end of the stop band, due to a reduction of the effective dynamic mass, after which the STL evolves towards the bare host structure performance [40, 42, 43]. Assouar et al. [44] analyzed the STL of an LRM plate with a 2D periodic array of rubber stub resonators. Hall et al. [45] measured the diffuse field STL of gypsum panels with beam-shaped resonators, in view of lightweight building structures. Van Belle et al. [43] measured the acoustic insulation of a PMMA LRM plate with add-on beam-shaped resonators and reported that resonator damping can improve the STL dip near the end of the stop band at the cost of a reduced peak STL. The potential of LRM sandwich panels with mass-spring or stubbed resonators for STL improvements was also theoretically investigated [41, 46]. In view of lightweight design integration, Claeys et al. [47] proposed a 3D printed LRM enclosure. By adding resonant structures in the hollow core of a periodic sandwich structure, a resonance-based bending wave stop band was obtained. This resulted in superior acoustic insulation in a targeted frequency range compared to a regular enclosure of the same mass (Fig. 4), although it also showed a reduced acoustic insulation in the frequency range after the stop band due to its reduced effective dynamic mass. Nevertheless, the potential of vibro-acoustic LRMs to combine lightweight and compact design with favourable noise and vibration attenuation in specifically targeted frequency ranges was thereby demonstrated.

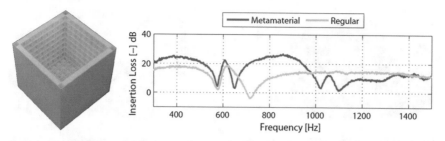

Fig. 4 Lightweight LRM enclosure (left) and comparison of the measured acoustic insertion loss of the LRM and of a regular enclosure of the same mass, showing a strongly improved acoustic insulation between 690–980 Hz (right) [47]

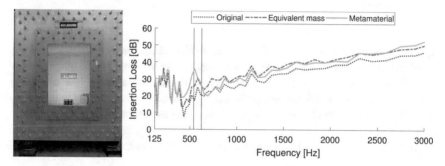

Fig. 5 LRM double panel sealing an acoustic cavity (left) and comparison of the measured acoustic insulation for the original double panel, the LRM double panel and a double panel of the same total mass (right), showing a strong improvement for the LRM panel around the stop band (vertical lines) tuned to the mass-air-mass resonance of the original double panel [42]

The applicability of LRMs to treat typical low-frequency noise and vibration problems is also increasingly explored. de Melo Filho et al. [42] investigated the use of the LRM concept to enhance the STL of an air-filled double panel at its mass-air-mass resonance frequency. A strong improvement of the acoustic insulation around the mass-air-mass resonance frequency range was measured due to the tuned stop band, outperforming the effect of regular homogeneous mass addition in the same frequency range (Fig. 5). de Melo Filho et al. [48] also brought LRMs closer to industrial realisations by exploiting a cheap and widely used production process. The potential was demonstrated of a twin sheet thermoformed LRM partition to reduce the interior booming noise in a closed cavity around a specific frequency range, for exterior acoustic excitation. Compared to the original panel, the LRM panel is lighter and has the same volume, but achieves an 8 dB reduction of the sound pressure level in the cavity for the targeted first acoustic cavity mode (Fig. 6).

In summary, vibro-acoustic LRMs have shown great potential to enable superior vibration and noise attenuation due to their resonance-based stop band behavior, albeit in specifically targeted and currently predominantly narrowband frequency ranges. The sub-wavelength nature of LRMs allows to achieve this with thinner and

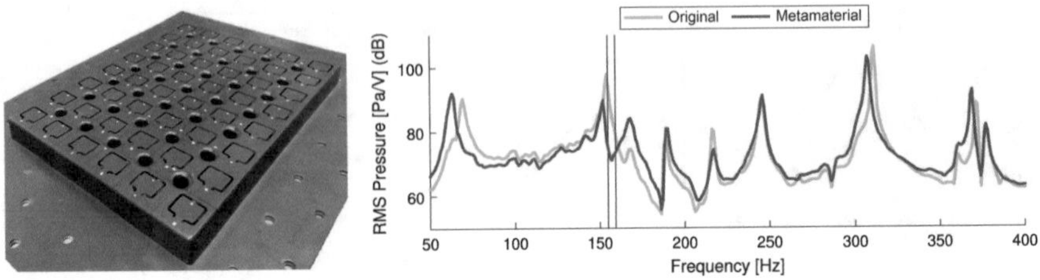

Fig. 6 Thermoformed LRM panel (left) and comparison of the measured sound pressure level in a resonant cavity sealed by the original and the LRM thermoformed panel for acoustic excitation outside the cavity (right), showing strong reduction of the targeted acoustic mode around the tuned stop band (vertical lines), followed by an overall slightly increased response after the targeted frequency range [48]

lighter vibro-acoustic solutions, yet potential performance trade-offs outside the targeted frequency range should be taken into account. Increasing evidence has already been presented of their applicability for typically hard-to-address low-frequency noise and vibration problems, integration in lightweight design as well as manufacturability. However, in view of bringing LRMs closer to becoming broadly applicable engineering solutions, achieving broadband vibro-acoustic performance is strongly desired, which is of particular interest in current LRM research [5, 49].

3   Vibro-Acoustic LRMs for Automotive NVH

Because of their appealing potential as thin and lightweight NVH solution, LRMs have attracted attention for NVH reduction in automotive applications. A variety of numerical and experimental assessments have recently been performed to investigate their potential for interior NVH reduction in targeted frequency ranges.

Wu et al. [50] numerically assessed the incorporation of the LRM concept in car floor panels, considering simplified mass-spring resonators. It was shown that the total body vibration levels could be sharply reduced and that also the interior noise level at the driver's ear could be reduced around a targeted frequency range. Similarly, Wu et al. [51] numerically assessed the application of LRMs to a car ceiling to achieve interior noise reduction, also showing potential to reduce interior noise levels at the driver's ear location in specific frequency ranges. Chang and Jung [52, 53] designed resonators which were applied to a car dash panel to induce stop band behavior. Using a cut-out mock-up front section of a body-in-white, the LRM concept was experimentally demonstrated in lab conditions to reduce dash panel vibration levels as well as radiated noise for a structural excitation, in three different stop band frequency ranges. Together with the authors of this chapter, Chang et al. [54] performed an initial investigation of the application of the LRM concept to the firewall of a car. The firewall was treated with patches of resonant structures, applied to identified regions of highest vibration (Fig. 7), and the thickness of the heavy layer

Fig. 7 Interior view of the exposed firewall of a vehicle, treated with added patches of resonators (left) and close up of a resonator unit (right) [54]

was reduced. This resulted in a weight reduction of 52% as compared to the original dash insulation pad. In-lab tests on a full vehicle without dashboard indicated the ability to slightly reduce vibrations and interior noise in targeted frequency ranges, while maintaining performance in non-targeted frequency ranges. The latter might originate from the additional air gaps which the resonator patches locally introduce.

The preceding works focus either on numerical assessments or on in-lab tests on simplified vehicle parts or under non-driving conditions. Recently, however, the authors of this chapter demonstrated the potential of the LRM concept to reduce interior noise levels in a targeted frequency range, in a real car and under real driving conditions [55]. This application is discussed in the following section.

4 LRM Application for Interior Noise Reduction

In this section, an LRM concept is applied to the rear shock towers of a commercial vehicle in view of reducing the noise inside the passenger compartment caused by vibrational energy, originating from the tire-road interaction, which flows into the vehicle body through the rear suspension and which causes body parts to vibrate and to radiate noise [56, 57]. This mainly narrowband NVH problem typically occurs in the low-frequency range and the dominant frequency depends on several mechanisms related to among others the tire dynamics, acoustic modes of the tires, the tire-road interaction, the suspension type and the driving conditions [58–62].

The considered car is a European sports utility vehicle (SUV), which has a MacPherson rear suspension and 235/60 R18 tires. Because of the type of suspension and acoustic tire mode, a high noise level in the passenger compartment results around 190 Hz. The SUV comes equipped with a tuned vibration absorber (TVA) on the rear shock towers (Fig. 8). These TVAs are designed to reduce the vibration energy entering the vehicle body through the rear suspension around 190 Hz, in order to reduce the noise level in the passenger compartment. This NVH solution currently adds 1.46 kg to each of the rear shock towers. The objective of the LRM application is to replace the TVAs in order to reduce mass without reducing NVH performance. In what follows, the design, integration and experimental validation of the LRM solution are discussed.

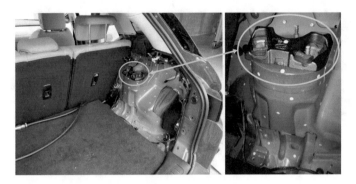

Fig. 8 TVA installed on top of the right rear shock tower of the vehicle

4.1 LRM Design

An LRM solution is designed, which will be added on the rear shock towers around the top mount. This location is chosen since it constitutes part of the transfer path for the vibrational energy traveling from the tires into the vehicle body. The targeted frequency range is 190 Hz, which has been identified through an experimental modal analysis of the current TVA solution. At this frequency, the TVAs have a dominant in-plane vibration orientation parallel to the top mount surface, which counteracts the vibrations of the shock tower excited through the top mount. The body panels constituting the shock towers can vibrate in both in- and out-of-plane direction. To efficiently reduce the energy propagation through the shock towers, the LRM solution hence needs to tackle both the in- and out-of-plane vibration propagation.

To achieve resonance-based stop band behavior, the LRM design needs to meet two conditions with respect to the targeted structural wave type(s) of interest (i) the net force exerted by the resonators on the host structure needs to be non-zero [63] and (ii) the resonators should be added onto the host structure on a sub-wavelength scale [26]. To comply with the first condition, a resonator is designed which consists of two cantilever beams, connected by a support, with an end point mass on each beam end (Fig. 9a). This design allows to have two low-frequency bending modes which can exert a net non-zero force on the host structure for both in- and out-of-plane vibrations [47], which can be straightforwardly tuned by changing the dimensions of the end point masses and the beams. For the current NVH problem, the resonators are designed to be effective around 190 Hz and to have the same resonance frequency for the first in-plane and out-of-plane bending mode (Fig. 9c). These resonators can be attached with the base of the support to the targeted body surfaces, while their spacing will be chosen to satisfy the second condition.

For the realization of the resonators, selective laser sintering was chosen, as it is a versatile production process which allows fast and accurate realization of complex resonator geometries (Fig. 9b). The properties of the used polyamide material are shown in Table 1. For the tuning of the resonator, Finite Element (FE) based modal

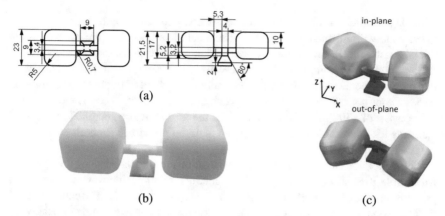

Fig. 9 (**a**) Resonator geometry (mm), (**b**) laser sintered resonator and (**c**) numerically predicted first in- and out-of-plane bending modes of the resonator at 193 Hz

Table 1 Material properties of polyamide and steel

	Young's modulus (GPa)	Density (kg/m³)	Poisson's ratio
Polyamide	2.05	965	0.4
Steel	210	7800	0.3

analysis is applied. Using an FE model consisting of 6807 linear solid elements, an initial design was proposed with a tuned resonance frequency around 190 Hz for a clamped resonator base. The resonance frequencies of 8 manufactured resonators were verified by measuring the vibration response of the end point masses with a Laser Doppler Vibrometer (Polytec PSV-500) [55]. The resulting resonance frequency of the manufactured resonators is 193 Hz (Fig. 9c), which is close enough to the targeted frequency range around 190 Hz. The geometrical dimensions of the resonators in the numerical model were updated in order for the numerically predicted resonance frequency to agree. The updated resonator model is next used in the prediction of the stop band behavior, which is explained in the following section.

4.1.1 Stop Band Prediction

To predict stop band behavior in LRMs, unit cell (UC) modeling is used. An undamped FE model of a single UC is combined with Bloch-Floquet periodic boundary conditions to calculate wave propagation in the assumed infinite periodic structure by means of dispersion curves [64]. Stop bands are found as frequency ranges in which no freely propagating wave solutions occur, which are known to correspond well to the frequency range of strong vibration attenuation in finite LRM counterparts.

In reality, the shock tower is an irregularly shaped and curved shell. To facilitate the stop band analysis by means of dispersion curves, a flat UC representing a structure with 2D periodicity is chosen to describe the LRM solution comprised of the shock tower plate with resonators (Fig. 10a). The steel plate host structure has a constant thickness of 2 mm, which is the average thickness of the shock tower panels around the top mounts. To comply with the sub-wavelength requirement and to efficiently make use of the available space, the resonators are added to the host structure with 65×25 mm periodicity, defining the eventual UC geometry. The FE model of the UC consists of 429 linear shell elements for the host structure and 6807 linear solid elements for the resonator. The considered material properties for both constituents are listed in Table 1.

The dispersion curves for free wave propagation are calculated along the so-called Irreducible Brillouin Contour (IBC) [10], obtained by imposing wave vectors along principal combinations of the periodicity lattice vectors d_1 and d_2. The coordinates of the corresponding imposed dimensionless propagation constants in reciprocal wave space are indicated by O, A, B, C, O (Fig. 10a). In the calculated dispersion diagram of the bare host structure UC three wave types are identified (Fig. 10b): out-of-plane bending (B) and in-plane shear (S) and longitudinal (L) wave propagation. The colouring indicates the out-of-plane nature of the waves in the host structure, with dark red indicating out-of-plane waves and dark blue in-plane waves. It is noted that, in reality, the complex geometry of the actual shock tower leads to a highly coupled nature of the waves, with waves converting to different wave types at junctions of flat or curved plates with different angles [65]. As described in the foregoing, the LRM design hence aims to target both in- and out-of-plane waves, in order to minimize the transmission of the vibration energy along the treated transfer path.

In the dispersion curves of the LRM UC (Fig. 10c), an omnidirectional stop band is achieved between 193.1 Hz and 224.1 Hz for the bending wave type (B). Furthermore, a directional stop band between 193 and 224 Hz is achieved for the shear wave type (S) along the d_1-direction and for the longitudinal wave type (L) along the d_2-direction. As a consequence, the proposed LRM design will allow to reduce the transmission of vibration energy along the treated transfer paths, whether it is carried by bending or, depending on the orientation, longitudinal or shear waves.

4.1.2 Surface Treatment Approach

As described before, the LRM solution has to be added on the rear shock towers around the top mount. To facilitate the manufacturing of the resonators and the installation on the vehicle, the resonators are designed and produced in patches which have a similar form as the rear shock tower geometries, which was achieved with the aid of a CAD model of the rear shock towers (Fig. 11). A total of four separate patches per shock tower has been added.

To determine the amount of resonators which have to be added on the shock towers, insights from a previous study by the authors on a conceptual shock tower were relied on [66]. This study revealed a minimum number of required resonators along a transfer path to effectively prevent energy from propagating through the structure. The size of the treated area was hence chosen as a compromise between occupying

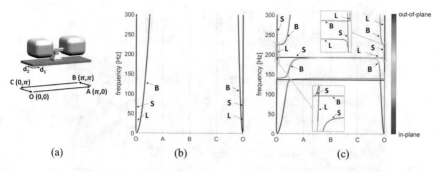

Fig. 10 (**a**) LRM UC consisting of the host structure and added resonator, with periodicity lattice vectors $\mathbf{d}_1$ and $\mathbf{d}_2$ and the imposed propagation constants along the IBC given by O, A, B, C, O, (**b**) dispersion diagram for the bare host structure UC and (**c**) the LRM UC, with (B), (S) and (L) indicating the different wave types

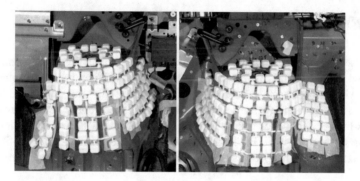

Fig. 11 Installed resonator patches on the rear shock towers. The left and right pictures represent the left and the right shock tower, respectively

the least space required and still hosting a minimum of five rows of resonators on the shock tower surface around the force input location. The chosen layout results in a total of 105 resonators, distributed over the rear shock towers, spaced on an as regular as possible grid corresponding to the 65 × 25 mm UC dimensions. Since the geometry of the rear shock towers is not symmetric, the right shock tower hosts three resonators more than the left shock tower.

To create the patches, the bases of the resonators are connected by a grid of rectangular beams with 1 × 4 mm cross section. These dimensions provide sufficient flexibility for the connections to ensure a good fit of the patches onto the shock tower, compensating possible geometrical variations due to manufacturing of the patches and geometrical discrepancies between the CAD model and the real shock tower geometries. The patches are connected to the shock tower with Loctite®406 contact adhesive. The eight patches together have a total mass of 1.52 kg, reducing the added mass by 48% as compared to the current NVH solution consisting of the two TVAs.

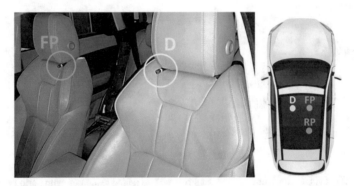

Fig. 12 Microphone positions for the driver (D), front passenger (FP) and rear passenger (RP) locations (car top view image courtesy of www.freepik.com)

4.2 Experimental Validation

In this section, the performance of the realized LRM solution is experimentally validated using in-vehicle measurements for on-road driving tests. An asphalt road profile with smooth surface was chosen for the tests. The tests were performed on dry public roads with low traffic such that constant speed could be maintained and no pass-by-noise influence from other vehicles occurred during the acquisitions. The used driving speed was 50 km/h, which was maintained constant within ± 5%. This speed is the maximum speed allowed on the public roads used in the driving tests.

4.2.1 Measurement Setup

Since an interior noise problem is of interest, the noise levels inside the passenger compartment are measured. Three microphones (PCB Model 378B02) are used, located at the driver right ear (D), front passenger left ear (FP) and rear right passenger left ear (RP) locations (Fig. 12). These precise locations are chosen in order to have an adequate distribution of the sensors to measure the noise perceived by the driver and passengers and to also minimize the influence of other noise sources, such as the windows, by measuring close to the longitudinal axis of the vehicle. The microphones are placed between the head restraint and the seat backrest. During the tests, the air temperature in the passenger compartment is set to 20°C and monitored to remain constant. The inflation pressure of the tires is 2.1 bar.

Three different configurations have been considered for the on-road tests, which are described in Table 2. The excitation level to which the vehicle is subjected during the driving tests is characterized using two tri-axial accelerometers (PCB Model 356A15) installed on the wheel knuckle of both rear wheels. This allows to verify that similar excitation levels are reached when testing the different configurations.

Table 2 Overview of the configurations for the on-road tests

Configuration	Description
Bare	No NVH treatment installed on the rear shock towers
TVA	TVAs installed on the rear shock towers
LRM	TVAs replaced by the LRM patches

As no input forces are measured, the Power Spectral Density (PSD) of the measured sound pressure levels (SPL) and of the accelerations in the three directions is evaluated, assuming steady-state harmonic response conditions for the tests. In order to obtain a converged spectrum, 150 averages were required per test and each test was repeated five times per configuration. For each sensor, the Root Mean Square (RMS) of the PSD across the number of measurements is calculated as:

$$RMS^i_{PSD_j}(\omega) = \sqrt{\frac{1}{N} \sum_{i=1}^{N} \mid PSD_{j,i}(\omega) \mid^2}, \tag{1}$$

with ω the angular frequency, N the total number of measurements, i the measurement number and j the quantity measured by the sensor (SPL or acceleration per direction). For the evaluation of the total acceleration from the three-axial accelerometers, an additional post-processing step is required following Eq. (1). The total RMS PSD acceleration for each accelerometer is calculated by taking the RMS of the RMS PSD accelerations along the three directions as follows:

$$RMS^{tot}_{PSD_{acc}}(\omega) = \sqrt{\sum_{i=1}^{3} \left(RMS^i_{PSD_{acc}}(\omega)\right)^2}, \tag{2}$$

with i indicating the accelerometer directions according the SAE standard [67].

4.2.2 Measurement Results

The three configurations from Table 2 are measured under real driving conditions and the results are analyzed in the frequency range of interest between 100 and 300 Hz, around the considered low-frequency NVH problem. First, the accelerometer measurements are investigated, to ensure consistency of the excitation for the different configurations. As shown in Fig. 13, the RMS PSD accelerations indicate that the excitation level for all configurations is similar. Two sharp peaks are also observed in the PSD acceleration. The highest acceleration peak occurs at 198 Hz for all configurations and originates from an acoustic tire resonance.

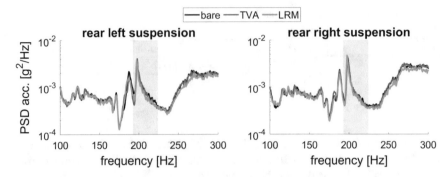

Fig. 13 RMS PSD acceleration at the rear left and right wheel knuckle for the three tested configurations. The predicted stop band is indicated by the gray shaded area

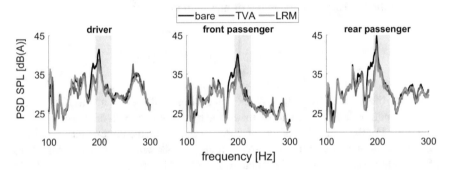

Fig. 14 PSD SPL for the three tested configurations at the three microphone locations. The predicted stop band is indicated by the gray shaded area

To assess the interior noise levels, the RMS PSD SPL is investigated (Fig. 14). The maximum standard deviation for the RMS PSD SPL of each configuration, averaged over the 100–300 Hz frequency range, lies within ±1.0 dB(A), which confirms the consistency of the different measurements of the same configuration. The SPL spectra are well correlated with the acceleration spectra, with the highest SPL peak emerging at 198 Hz. Especially for the bare configuration, this peak is very outspoken, which clearly highlights the low-frequency NVH problem of interest: high interior noise levels are caused by vibrational energy, generated by tire-road interaction, which flows into the vehicle body through the rear suspension.

When inspecting the interior noise levels for the TVA and LRM configurations, a clear reduction of the strong SPL peak at 198 Hz is obtained at all measurement locations. In both cases, a substantial improvement of the SPL and thus interior noise levels with respect to the bare configuration results between 185 and 210 Hz. The highest achieved SPL reduction in the passenger compartment with the LRM solution occurs at frequencies close to the lower bound of the predicted stop band, which is typically close to the tuned resonance frequency of the resonator. At the highest SPL peak at 198 Hz, the LRM even outperforms the TVA configuration, while adding

Table 3 SPL values at 198 Hz for the three configurations at the three locations. The mean value and standard deviation of the 5 measurements is shown

Configuration	D [dB(A)]	FP [dB(A)]	RP [dB(A)]
Bare	41.4 ± 0.8	40.3 ± 0.8	44.7 ± 0.9
TVA	38.8 ± 0.8	38.3 ± 0.7	42.6 ± 0.7
LRM	37.8 ± 0.7	36.6 ± 0.7	40.9 ± 0.8

48% less mass. This is also shown in Table 3, which summarizes the SPL results at 198 Hz for the three configurations. Before 185 Hz and after 210 Hz, the results for the three configurations are comparable.

Based on the presented results, the potential of the LRMs for improved interior NVH performance has clearly been verified for on-road driving conditions in a real vehicle. Apart from showing a similar or better performance as compared to the current NVH solution based on two TVAs, the LRM concept is also shown to allow a considerable reduction of the required mass addition. With this successful application, LRMs are brought closer to becoming industrially applicable lightweight NVH solutions which provide lightweight replacements of current NVH treatments to tackle the hard-to-address low-frequency range.

5 Conclusions

In view of reducing emissions and improving quality and comfort in vehicles, automotive companies seek for lightweight NVH solutions. However, lightweight structures typically lead to worsened NVH performance. A variety of novel NVH solutions emerged, which try to combine lightweight design with improved NVH performance, among which vibro-acoustic LRMs have shown great potential.

In this chapter, an overview was given of the emergence of vibro-acoustic LRMs. These LRMs exhibit stop bands, arising from sub-wavelength addition of resonant structures to a host structure, which enables specifically targeted frequency ranges of strong vibration and noise reduction, also for the hard-to-address low-frequency range. Numerical and experimental demonstrators of increasing complexity have already proven the potential of these LRMs to design light and compact structures with improved NVH performance, albeit currently in predominantly narrowband frequency ranges and at the cost of possible performance reducing side effects outside the targeted frequency ranges. Very recently, LRMs have drawn attention for automotive applications and first promising numerical and in-lab experimental demonstrations of their use for interior NVH improvements have been presented.

The last part of this chapter presented the design and application of an LRM solution for a real vehicle, in order to improve the interior noise levels in the passenger compartment around 200 Hz. Resonators were designed and realized in patches to treat the rear shock towers, to prevent vibrations caused by an acoustic tire mode from entering the passenger compartment. On-road tests in real driving conditions demonstrated a similar or better performance of the LRM solution as compared to the currently used NVH solution, while reducing the added mass with 48%. This demonstration brings LRMs closer to becoming an applicable lightweight NVH solution for the automotive industry.

Acknowledgments The Research Fund KU Leuven is gratefully acknowledged for its support. This research was partially supported by Flanders Make, the Strategic Research Center for the manufacturing industry. The European Commission is gratefully acknowledged for their support of the PBNv2 (GA 721615), SMARTANSWER (GA 722401) and VIPER (GA 675441) research projects. The research of E. Deckers is funded by a grant from the Research Foundation – Flanders (FWO). M. Clasing acknowledges CONICyT for a Becas Chile scholarship.

References

1. Taub AI, Krajewski PE, Luo AA, Owens JN (2007) The evolution of technology for materials processing over the last 50 years: the automotive example. JOM 59(2):48–57
2. World Health Organization (2018) Environmental noise guidelines for the European region
3. Goetchius G (2011) Leading the charge–the future of electric vehicle noise control. Sound Vibr 45(4):5–8
4. Hussein MI, Leamy MJ, Ruzzene M (2014) Dynamics of phononic materials and structures: historical origins, recent progress, and future outlook. Appl Mech Rev 66(4):040802
5. Ma G, Sheng P (2016) Acoustic metamaterials: from local resonances to broad horizons. Sci Adv 2(2):e1501595
6. Veselago VG (1968) The electrodynamics of substances with simultaneously negative values of ϵ and μ. Sov Phys Uspekhi 10(4):509–514
7. Christensen J, Kadic M, Kraft O, Wegener M (2015) Vibrant times for mechanical metamaterials. MRS Commun 5(3):453–462
8. Liu Z, Zhang X, Mao Y, Zhu YY, Yang Z, Chan CT, Sheng P (2000) Locally resonant sonic materials. Science 289(5485):1734–1736
9. Fok L, Ambati M, Zhang X (2008) Acoustic metamaterials. MRS Bull 33(10):931–934
10. Brillouin L (1946) Wave propagation in periodic structures. Int Ser Phys
11. Ho KM, Chan CT, Soukoulis CM (1990) Existence of a photonic gap in periodic dielectric structures. Phys Rev Lett 65(25):3152–3155
12. Yablonovitch E, Gmitter TJ, Leung KM (1991) Photonic band structure: the face-centered-cubic case employing nonspherical atoms. Phys Rev Lett 67(17):2295–2298
13. Kushwaha MS, Halevi P, Dobrzynski L, Djafari-Rouhani B (1993) Acoustic band structure of periodic elastic composites. Phys Rev Lett 71(13):2022–2025
14. Sigalas MM, Economou EN (1992) Elastic and acoustic wave band structure. J Sound Vibr 158(2):377–382
15. Kittel C (2010) Introduction to solid state physics
16. Pendry JB, Holden AJ, Robbins DJ, Stewart WJ (1999) Magnetism from conductors and enhanced nonlinear phenomena. IEEE Trans Microwave Theory Tech 47(11):2075–2084
17. Li J, Chan CT (2004) Double-negative acoustic metamaterial. Phys Rev E 70(5):055602
18. Goffaux C, Sánchez-Dehesa J, Yeyati AL, Lambin P, Khelif A, Vasseur JO, Djafari-Rouhani B (2002) Evidence of Fano-like interference phenomena in locally resonant materials. Phys Rev Lett 88(22):225–502
19. Liu Z, Chan CT, Sheng P (2005) Analytic model of phononic crystals with local resonances. Phys Rev B Condens Matter Mater Phys 71(1):1–8

20. Calius EP, Bremaud X, Smith B, Hall A (2009) Negative mass sound shielding structures: early results. Physica Status Solidi 246(9):2089–2097
21. Ho KM, Cheng CK, Yang Z, Zhang XX, Sheng P (2003) Broadband locally resonant sonic shields. Appl Phys Lett 83(26):5566–5568
22. Hsu JC, Wu TT (2007) Lamb waves in binary locally resonant phononic plates with two-dimensional lattices. Appl Phys Lett 90(20):201904
23. Krushynska A, Kouznetsova V, Geers M (2014) Towards optimal design of locally resonant acoustic metamaterials. J Mechan Phys Solids 71:179–196
24. Xiao W, Zeng GW, Cheng YS (2008) Flexural vibration band gaps in a thin plate containing a periodic array of hemmed discs. Appl Acoust 69(3):255–261
25. Xiao Y, Wen J, Wen X (2012) Flexural wave band gaps in locally resonant thin plates with periodically attached springmass resonators. J Phys D Appl Phys 45(19):195401
26. Claeys C, Vergote K, Sas P, Desmet W (2013) On the potential of tuned resonators to obtain low-frequency vibrational stop bands in periodic panels. J Sound Vibr 332(6):1418–1436
27. Wu TT, Huang ZG, Tsai TC, Wu TC (2008) Evidence of complete band gap and resonances in a plate with periodic stubbed surface. Appl Phys Lett 93(11):98–101
28. Oudich M, Senesi M, Assouar MB, Ruzenne M, Sun JH, Vincent B, Hou Z, Wu TT (2011) Experimental evidence of locally resonant sonic band gap in two-dimensional phononic stubbed plates. Phys Rev B 84(16):165136
29. Assouar B, Senesi M, Oudich M, Ruzzene M, Hou Z (2012) Broadband plate-type acoustic metamaterial for low-frequency sound attenuation. Appl Phys Lett 101(17):173–505
30. Xiao Y, Wen J, Huang L, Wen X (2014) Analysis and experimental realization of locally resonant phononic plates carrying a periodic array of beam-like resonators. J Phys D Appl Phys 47(4):045307
31. Van Belle L, Claeys C, Deckers E, Desmet W (2017) On the impact of damping on the dispersion curves of a locally resonant metamaterial: modelling and experimental validation. J Sound Vibr 409:1–23
32. Claeys C, Vivolo M, Sas P, Desmet W (2012) Study of honeycomb panels with local cell resonators to obtain low-frequency vibrational stopbands. In: Proceedings of DYNACOMP 2012. Arcachon, France
33. Liu L, Hussein MI (2012) Wave motion in periodic flexural beams and characterization of the transition between Bragg scattering and local resonance. J Appl Mechan 79(1):011003
34. Xiao Y, Wen J, Wang G, Wen X (2013) Theoretical and experimental study of locally resonant and Bragg band gaps in flexural beams carrying periodic arrays of beam-like resonators. J Vibr Acoust 135(4):041006
35. Claeys C, de Melo Filho NGR, Van Belle L, Deckers E, Desmet W (2017) Design and validation of metamaterials for multiple structural stop bands in waveguides. Ext Mechan Lett 12:7–22
36. Nateghi A, Sangiuliano L, Claeys C, Deckers E, Pluymers B, Desmet W (2019) Design and experimental validation of a metamaterial solution for improved noise and vibration behavior of pipes. J Sound Vibr 455:96–117
37. Claeys C, Sas P, Desmet W (2014) On the acoustic radiation efficiency of local resonance based stop band materials. J Sound Vibr 333(14):3203–3213
38. Li P, Yao S, Zhou X, Huang G, Hu G (2014) Effective medium theory of thin-plate acoustic metamaterials. J Acoustic Soc Am 135(4):1844–1852
39. Oudich M, Zhou X, Assouar MB (2014) General analytical approach for sound transmission loss analysis through a thick metamaterial plate. J Appl Phys 116(19):193509
40. Xiao Y, Wen J, Wen X (2012) Sound transmission loss of metamaterial-based thin plates with multiple subwavelength arrays of attached resonators. J Sound Vibr 331(25):5408–5423
41. Song Y, Feng L, Wen J, Yu D, Wen X (2015) Reduction of the sound transmission of a periodic sandwich plate using the stop band concept. Compos Struct 128:428–436
42. de Melo Filho NGR, Van Belle L, Claeys C, Deckers E, Desmet W (2019) Dynamic mass based sound transmission loss prediction of vibro-acoustic metamaterial double panels applied to the mass-air-mass resonance. J Sound Vibr 442:28–44

43. Van Belle L, Claeys C, Deckers E, Desmet W (2019) The impact of damping on the sound transmission loss of locally resonant metamaterial plates. J Sound Vibr 461:114909
44. Assouar B, Oudich M, Zhou X (2016) Métamatériaux acoustiques pour l'isolation sonique. Comptes Rendus Physique 17(5):524–532
45. Hall A, Dodd G, Calius E (2017) Diffuse field measurements of locally resonant partitions. In: Proceedings of ACOUSTICS 2017
46. Li J, Li S (2017) Sound transmission through metamaterial-based double-panel structures with poroelastic cores. Acta Acustica United with Acustica 103(5):869–884
47. Claeys C, Deckers E, Pluymers B, Desmet W (2016) A lightweight vibro-acoustic metamaterial demonstrator: numerical and experimental investigation. Mechan Syst Signal Process 70–71:853–880
48. de Melo Filho NGR, Claeys C, Deckers E, Desmet W (2019) Realisation of a thermoformed vibro-acoustic metamaterial for increased STL in acoustic resonance driven environments. Appl Acoust 156:78–82
49. Ge H, Yang M, Ma C, Lu MH, Chen YF, Fang N, Sheng P (2018) Breaking the barriers: advances in acoustic functional materials. Natl Sci Rev 5(2):159–182
50. Wu X, Sun L, Zuo S, Liu P, Huang H (2019) Vibration reduction of car body based on 2D dual-base locally resonant phononic crystal. Appl Acoust 151:1–9
51. Wu X, Zhang M, Zuo S, Huang H, Wu H (2019) An investigation on interior noise reduction using 2D locally resonant phononic crystal with point defect on car ceiling. J Vibr Control 25(2):386–396
52. Chang KJ, Jung J, Kim HG, Choi DR, Wang S (2018) An application of acoustic metamaterial for reducing noise transfer through car body panels. SAE Technical Paper 2018-01-1566
53. Jung J, Kim HG, Goo S, Chang KJ, Wang S (2019) Realisation of a locally resonant metamaterial on the automobile panel structure to reduce noise radiation. Mechan Syst Signal Process 122:206–231
54. Chang KJ, de Melo Filho NGR, Van Belle L, Claeys C, Desmet W (2019) A study on the application of locally resonant acoustic metamaterial for reducing a vehicle's engine noise. In: Proceedings of Internoise 2019
55. Sangiuliano L, Claeys C, Deckers E, De Smet J, Pluymers B, Desmet W (2019) Reducing vehicle interior nvh by means of locally resonant metamaterial patches on rear shock towers. SAE Technical Paper 2019-01-1502
56. Douville H, Masson P, Berry A (2006) On-resonance transmissibility methodology for quantifying the structure-borne road noise of an automotive suspension assembly. Appl Acoust 67(4):358–382
57. Kindt P, Berckmans D, De Coninck F, Sas P, Desmet W (2009) Experimental analysis of the structure-borne tyre/road noise due to road discontinuities. Mechan Syst Signal Process 23(8):2557–2574
58. Hartleip LG, Roggenkamp TJ (2005) Case study-experimental determination of airborne and structure-borne road noise spectral content on passenger vehicles. SAE Technical Paper 2005-01-2522
59. Kido I, Nakamura A, Hayashi T, Asai M (1999) Suspension vibration analysis for road noise using finite element model. SAE Technical Paper 1999-01-1788
60. Kim G, Holland K, Lalor N (1997) Identification of the airborne component of tyre-induced vehicle interior noise. Appl Acoust 51(2):141–156
61. Sakata T, Morimura H, Ide H (1990) Effects of tire cavity resonance on vehicle road noise. Tire Sci Technol 18(2):68–79
62. Tatlow J, Ballatore M (2017) Road noise input identification for vehicle interior noise by multi-reference transfer path analysis. Proc Eng 199:3296–3301
63. Wang G, Wen X, Wen J, Shao L, Liu Y (2004) Two-dimensional locally resonant phononic crystals with binary structures. Phys Rev Lett 93(15):154302
64. Mace BR, Manconi E (2008) Modelling wave propagation in two-dimensional structures using finite element analysis. J Sound Vibr 318(4–5):884–902

65. Cremer L, Heckl M (2013) Structure-borne sound: structural vibrations and sound radiation at audio frequencies. Springer, Berlin
66. Sangiuliano L, Claeys C, Deckers E, Pluymers B, Desmet W (2018) Force isolation by locally resonant metamaterials to reduce NVH. SAE Technical Paper 2018-01-1544
67. SAE: SAE J670, Vehicle dynamics terminology

Active Sound Control in the Automotive Interior

Jordan Cheer

Abstract Active sound control has seen extensive use in a variety of industries and has been a potential tool for noise and vibration control engineers in the automotive industry since the late 1980s. However, it has only relatively recently begun to be more widely adopted within the automotive sector. This is in part due to a reduction in the cost of implementation, but also due to changes in vehicle powertrain systems and the resulting NVH challenges, as well as the evolving demands of consumers. For example, active sound control technology can be used to not only reduce noise in the automotive environment, but to enhance the driving experience through the manipulation of powertrain noise or the provision of advanced infotainment features. Resultingly, active sound control systems form a potentially integral part in the design of future automotive interiors. This chapter reviews the development and state-of-the-art in active sound control systems within the automotive sector, including both noise reduction and sound reproduction technologies that can come together to enable the refinement of future interiors.

1 Introduction

Active sound control is the process of manipulating a sound field through the use of electroacoustic transducers, or typically loudspeakers. Perhaps the most obvious application is where a sound field is generated by the loudspeakers in order to destructively interfere with a primary sound field and thus attenuate an unwanted noise source; this is often referred to as Active Noise Control (ANC) [1]. More broadly, however, active sound control also encompasses the generation of wanted sound fields [2], as in the case of spatial audio reproduction [3] and zonal sound control [4]. Additionally, these concepts have also been combined, where rather than cancelling an unwanted noise source, a secondary sound field is generated to manipulate the

J. Cheer (✉)
Institute of Sound and Vibration Research, University of Southampton, Southampton SO17 1BJ, UK
e-mail: j.cheer@soton.ac.uk

© The Author(s), under exclusive license to Springer Nature Switzerland AG 2021
A. Fuchs and B. Brandstätter (eds.), *Future Interior Concepts*,
Automotive Engineering : Simulation and Validation Methods,
https://doi.org/10.1007/978-3-030-51044-2_3

primary sound field to produce a desired response, usually with improved sound quality [5].

In all of the above cases, it is necessary for the control system to generate a sound field with both accurately controlled spatial and temporal properties [6]. In order to achieve accurate spatial control over a large region requires a potentially large number of loudspeakers, and this can be physically related to the decrease in the acoustic wavelength with increasing frequency. This places a limit on the upper frequency at which active sound control can generally be achieved and these physical limits have been extensively discussed in the literature [1]. While spatial matching is limited by the physical acoustics, temporal matching is generally limited by delays in the signal processing system used in the controller. These can be somewhat limited by operating at high sample rates using high performance Digital Signal Processors (DSPs), however, limiting delays also occur due to the physical design of the control system. These limits have also been extensively investigated and are covered in a number of key texts [7, 8].

Active control of sound has an extended history with some early incarnations being proposed in the 1930s [9], but significant practical implementations were not realised until the late 1980s and 1990s following the increased availability and reduced cost of DSP [7]. This development boom saw researchers tackle a variety of practical noise control problems using active technologies. These include noise cancelling headphones [10], the control of propeller induced noise in aircraft [11, 12], the control of noise in helicopters [13] and the control of both engine [14] and road noise [15] in the automotive environment. Despite the successful demonstration of active control technologies in a variety of practical applications, aside from the commercial success of active noise cancelling headphones, its adoption has until recently been largely limited to less cost sensitive applications, such as the aircraft industry.

In the automotive sector, despite practical demonstrations of both engine [14] and road [15] noise control systems in the late 1980s and early 1990s respectively, and a small number of early commercial systems, such as the system implemented by Nissan in their Bluebird [16], widespread adoption has been limited. This was largely due to the cost of implementing these systems, which was initially exacerbated by a lack of integration between the common components required by both the car audio system and the control system. However, it can also be linked to people's expectations of active control systems and the progression of the technology through the 'trough of disillusionment', described by Gartner's Hype Cycle [17]. More recently, however, it appears that the expectations of active sound control systems have become more aligned with their limitations and their application in the automotive industry has increased, with systems in production that not only aim to reduce sound, but that enable active sound design [18]. Beyond NVH related challenges, active sound control technology has also been utilised to provide car cabin occupants with enhanced performance and functionality via the infotainment system. This includes improvements to the stereo imaging problems caused by off-centre listening positions [19], the provision of spatial audio reproduction capabilities [20, 21] and the generation of personal listening zones [22]. These aspects of future vehicle interiors are also becoming of increased importance as autonomous vehicles become a reality [23, 24].

This chapter will review the development of active sound control systems in the automotive environment, including systems for the reduction, manipulation and generation of controlled sound fields, all of which aim to improve the user experience. The physical limitations of ANC systems will first be discussed, with a consideration of both global and local control strategies. A review of both research and commercial implementations of ANC in the automotive environment will then be presented and future aspects discussed. Subsequently, developments in the field of active control for sound generation, or reproduction in the automotive environment will be reviewed and the links between these two areas will be made.

2 Physical Limitations of Active Noise Control

ANC uses an array of actuators, or typically loudspeakers, to control a primary sound field via the generation of a secondary sound field. Broadly speaking, this can either be realised on a global scale, with control being achieved throughout an environment, or on a local scale, with control focused over a particular region of space. These two approaches both have advantages and disadvantages that relate to both practical limitations in performance and cost of implementation. In the following sections, the basis of global and local active noise control strategies will be reviewed.

2.1 Global Active Control

Global active control of noise in an enclosure, such as a car cabin, aims to minimise the sound pressure level throughout the enclosure. This is usually achieved in theory by minimising the total acoustic potential energy within the enclosure, although in practice this has to be approximated by minimising the sum of the squared pressures measured at a number of error microphone locations. If the control system is implemented as a feedforward controller, as shown in Fig. 1, and consists of L error microphones located in the enclosure, M control loudspeakers and K reference signals, the vector of error signals, $\mathbf{e}$, at a single frequency can be expressed as

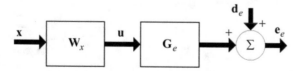

Fig. 1 Multichannel feedforward controller, with reference signals x, disturbance signals d, error signals e, control signals u, control filter W and plant response G

$$\mathbf{e} = \mathbf{d} + \mathbf{Gu} = \mathbf{d} + \mathbf{GWx}$$

where $\mathbf{d}$ is the vector of disturbance signals measured at the error microphones without control, $\mathbf{G}$ is the $(L \times M)$ matrix of transfer responses between the inputs to the control loudspeakers and the pressures measured at the error microphones, $\mathbf{u}$ is the vector of M control signals which drive the control loudspeakers, $\mathbf{x}$ is the vector of reference signals and $\mathbf{W}$ is the control filter matrix. In general, global active control aims to minimise the sum of the squared error signals, which gives the cost function to be minimised at a single frequency as

$$J = \mathbf{e}^H \mathbf{e} = \mathbf{u}^H \mathbf{G}^H \mathbf{Gu} + \mathbf{u}^H \mathbf{G}^H \mathbf{d} + \mathbf{d}^H \mathbf{Gu} + \mathbf{d}^H \mathbf{d}.$$

The optimal vector of control signals which minimises this cost function is then given by

$$\mathbf{u}_{\text{opt}} = -\left[\mathbf{G}^H \mathbf{G}\right]^{-1} \mathbf{G}^H \mathbf{d}.$$

Although the optimal solution assumes that disturbance signals are available in advance and does not represent the need for real-time adaptation required in practical applications in the automotive environment, it is useful for providing physical insight into the limits of ANC. Practical implementation would generally require the use of an adaptive algorithm, such as the widely employed Filtered-reference Least Mean Squares (FxLMS) algorithm [25].

The physical limitations of global control within a car cabin sized enclosure can be understood by considering the attenuation of an unwanted primary sound field within a rectangular, car cabin sized enclosure. Figure 2 shows the total acoustic potential energy in the rectangular enclosure before control and after control using either a single secondary source or 8 secondary sources, all located in the corners of

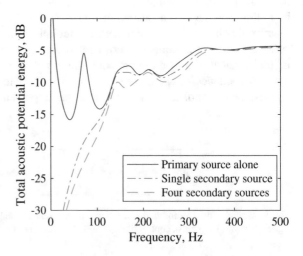

Fig. 2 The total acoustic potential energy in a three-dimensional rectangular enclosure before and after control using a single control source located in one corner and eight secondary control sources located in the corners of the enclosure

the enclosure. From this plot it can be seen that the single secondary source is able to achieve significant control at the compliant and first longitudinal mode at 72 Hz, but is unable to achieve significant attenuation at higher frequencies. This is because the single source is not able to effectively couple into the higher order modes of the enclosure. To achieve control of the higher order modes requires an increase in the number of secondary sources and it can be seen from the results presented in Fig. 2 that by increasing the number of sources to eight, and thus coupling into an increased number of modes, the bandwidth over which global control is achieved is increased to around 240 Hz. To extend the bandwidth further requires a significant increase in the number of secondary sources, since the number of modes in a three-dimensional enclosure increases approximately with the cube of the excitation frequency.

2.2 Local Active Noise Control

To overcome the upper frequency limits of global ANC imposed by the physical acoustics and thus increase the effective bandwidth of active control, there has been significant research interest in using local ANC to create zones of quiet around a listener's head using loudspeakers mounted in the headrest, for example [26, 27]. The performance of local ANC systems is dependent on both the geometry of the head and headrest [28], but also on the spatial properties of the sound field [29]. However, a useful rule of thumb based on the performance of a single-input, single-output (SISO) local active control system in a diffuse sound field states that the size of the zone of quiet within which 10 dB of attenuation is achieved is approximately one tenth of the acoustic wavelength [30]. Figure 3 shows the 10 dB zone of quiet achieved by a SISO local ANC system in a diffuse field at different frequencies. In this case the error sensor is located at (0, 0) m and the control source is located at

Fig. 3 The zone within which 10 dB of attenuation is achieved by a SISO local ANC system at various frequencies in a diffuse primary sound field, simulated according to the method outlined in [29]

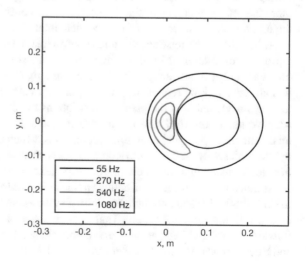

(0.1, 0) m. From this plot it can be seen that the size of the zone of quiet decreases with increasing frequency and has a diameter of around 6 cm at 540 Hz. Although local ANC thus offers a significant bandwidth extension over global active control, the performance will clearly be sensitive to the movement of the listener at higher frequencies where the zone of quiet is small. To extend the bandwidth of control further, a number of researchers have investigated methods of moving the zone of quiet with the head of the listener [31–35]. One such method applied to the road noise control problem will be discussed in further detail below.

3   Automotive Applications of Active Noise Control

This section will provide an overview of various ANC systems that have been implemented within the automotive environment. This includes control systems for both engine and road noise control, as well as global and local control strategies.

3.1   Active Control of Engine Noise

Despite the need for greener modes of transport, plug-in electric vehicles accounted for just 1% of global auto sales in 2015 [36] and although this is forecast to increase to around 60% by 2030, bolstered by the backing of global governments, it is clear that the Internal Combustion Engine (ICE) will remain dominant for a considerable period of time. This is particularly true with the availability of technological developments that increase the efficiency of the ICE such as: hybrid cars, which at least partially use a traditional ICE; variable cylinder management, which automatically deactivates some of the engine's cylinders according to the driving conditions; or simply smaller more fuel-efficient engine designs. Therefore, the control of the noise generated by the ICE remains of interest for future vehicle interiors.

Although passive noise and vibration control treatments for the reduction of engine noise are well-established and new technologies are still being developed [37], ANC offers a number of advantages. The first commonly cited advantage of ANC is that it is effective at low frequencies, where passive treatments become ineffective due to long wavelength and the practical restrictions on size and weight. However, the adaptive nature of ANC also means that it is able to provide an effective control treatment under changing operational conditions, which is particularly important for hybrid vehicles, or for those with technologies such as variable cylinder management in order to maintain a consistent driving experience [18].

As noted in the introduction, active engine noise control was first demonstrated in the late 1980s [14] and although a variety of systems have since been proposed, they largely follow a similar structure. That is, engine ANC is implemented using a feed-forward control architecture, where a reference signal is provided by a tachometer, and microphones and loudspeakers positioned in the car cabin provide the error signals and control sources respectively. This form of system can be readily integrated with modern car audio systems and, by utilizing the audio DSP, can be relatively cost efficient. As a result, a number of manufacturers have implemented such systems into production vehicles.

To demonstrate the performance limitations of an engine ANC system, Fig. 4 shows the performance of a system implemented in a small city car with a 2-cylinder engine. The system utilized a tachometer reference signal, 8 error microphones mounted at the four headrests, 4 low-frequency loudspeakers and the FxLMS adaptive control algorithm [38]. The plot in Fig. 4 shows the attenuation in the sum of the squared pressures measured at 16 microphones at the first engine order during an engine run-up. These results show significant levels of attenuation at low engine speeds, which generate low-frequency excitation, but the decrease in performance at higher engine speeds demonstrates the limits of global ANC.

Aside from noise cancelling, active control has also been used in the automotive environment to selectively attenuate or enhance features of the engine sound, with the objective of tailoring the driver experience. This is achieved using adaptive algorithms such as the active noise equalizer [39] or sound profiling algorithms [40, 41] that are based on modifications of the FxLMS algorithm. This technology has been utilized by automotive manufacturers to obtain a desired sound quality that is perhaps consistent with their brands [42], but also to cope with the complexities of hybrid or variable cylinder management ICEs [18]. In [42], for example, Kobayashi et al. present an active sound quality control system applied to a mass production vehicle. This system combines an active engine noise cancellation system to reduce engine boom, with a system to control the engine sound quality during acceleration. The system operates in its cancellation mode at lower engine speeds, where the vehicle is either idling or cruising, whilst at higher engine speeds, which occur during acceleration, the active sound quality control system aims to synthesize a "sporty sound". This is achieved

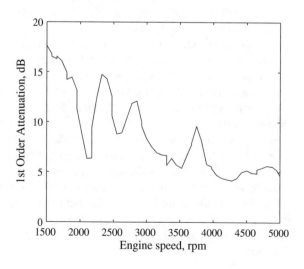

Fig. 4 The attenuation in the sum of the squared pressures at 16 microphones measured at the first engine order during an engine run-up [38]

by adaptively modifying the level of the individual engine orders with a dependency on the engine speed [42].

3.2   Active Control of Road Noise

Active control of road noise is a more challenging problem than active engine noise control due its complex and unpredictable nature. In the case of engine noise control, it is relatively straightforward to obtain a reference signal that is highly coherent with the sound field produced in the vehicle cabin due to the engine. For road noise control, however, the selection of reference signals that provide a time-advanced and coherent relationship to the contribution of road noise to the in-cabin sound field is much less straightforward. Despite these challenges, the first active road noise control systems were demonstrated in the early 1990s using similar error microphone and control loudspeaker configurations to the global engine noise control strategies [15, 43]. These systems, and subsequent developments [44], typically use reference signals obtained by direct measurement of the vibration induced by road excitation. For example, in [44], reference signals are provided by accelerometers mounted on the front and rear suspension systems. Due to the multiple propagation paths for road-tire induced vibration to reach the interior of the vehicle cabin, multiple reference signals are required to provide sufficient coherence to obtain useful levels of road noise control. This means that road noise control systems are costlier to implement than engine noise control systems, due to the need for additional sensors, wiring and a higher demand on the DSP. Despite this, active road noise control is an important technology to allow lightweight vehicle design, which tends to result in an increase in the low frequency broadband interior noise. This is particularly critical in electric vehicles where the masking noise provided by the ICE has been removed [45].

An alternative to employing multiple accelerometer reference signals for the real-isation of an active road noise control system is to use microphone reference sensors located in the vehicle cabin [46]. This may be a more cost-effective system design, as it reduces the need for the robust accelerometers positioned on the exterior of the vehicle [46]. Figure 5 shows the performance of this type of road noise control system implemented in a small city car using the four low-frequency car audio loudspeakers, four headrest error microphones and four additional microphone sensors mounted on the floor of the cabin. Figure 5 shows the sum of the squared pressures measured at the head-rest error microphones and it can be seen that the broadband peak in the road noise between 80 and 200 Hz has been attenuated by the feedforward controller and a maximum attenuation of 8 dB has been achieved at 115 Hz. At higher frequencies it can be seen that the performance of the road noise controller is rather limited. This can again be linked to the increased number of acoustic modes being excited, but in the case of the road noise controller can also be related to the low multiple-coherence between the reference microphone signals and the error microphone signals at higher frequencies.

Fig. 5 The sum of the squared pressures measured at the headrest error microphones in a small city car driven at 50 kph on a pave road surface before control (thick black line) and after control using a feedforward controller (thin black line) [46]

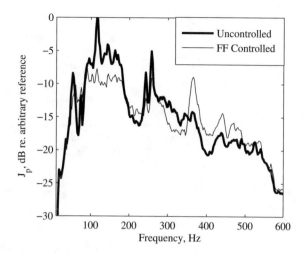

The performance of the road noise control system shown in Fig. 5 is largely consistent with the performance of the various systems presented in the open literature. From these results it is clear that although such global road noise control systems may be able to reduce low frequency boom, they are not able to provide control over an extended bandwidth without significantly increasing the number of control loudspeakers and potentially reference sensors. To overcome this limitation and extend the high frequency limit of active road noise control systems, local ANC with headtracking has been proposed [47].

Figure 6 shows an overview of the local ANC system with headtracking, as initially described in [33]. The in-car system, described in [47], uses two loudspeakers mounted in the car seat headrest to achieve control at the ears of the seat occupant. The error signals at the ears were estimated using the remote microphone method [29, 32] and the signals from 16 microphones positioned in the car cabin, as detailed in [47]. The reference signals for the feedforward controller were provided by 8 sensors located around the four wheels of the car and these reference sensors were carefully selected to provide both time-advance and a high level of multiple-coherence between the reference and error signals. The location of the occupant's head was tracked using a commercial headtracking system and the positional information was used to update the responses used in the controller and the estimation filters used in the remote microphone algorithm. As described in [47], these responses and filters must be identified during a preliminary calibration phase, but can be designed to be robust to practical uncertainties. Figure 7 shows the performance of the local ANC system implemented in a large Sports Utility Vehicle (SUV) when it is driven at 50 kph on a rough road surface. The results presented in this figure show the attenuation achieved at the left and right ears, and from these results it is clear that the control system is able to achieve attenuation in the road noise up to around 1 kHz. This system thus offers a significant performance advantage over both global active control systems and local ANC systems without headtracking, which are generally

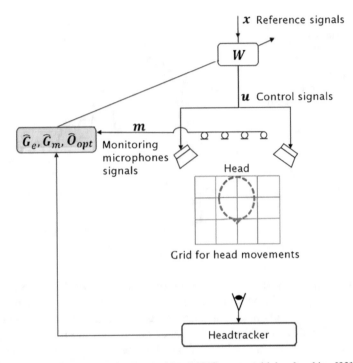

Fig. 6 Overview of the adaptive feedforward local ANC system with headtracking [33]

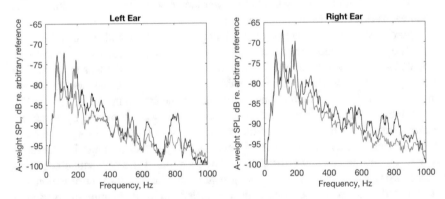

Fig. 7 The A-weighted sound pressure level measured at the left and right ears of the dummy head before control (black lines) and after control (red lines) using the feedforward local ANC system using remote sensing, with 8 reference signals, 16 monitoring microphones, 2 control loudspeakers and 2 error signals when the SUV is driven at 80 kph over a rough road surface [47]

limited to frequencies of around 500 Hz or so due to the size of the zone of quiet and realistic head movements.

Although the local ANC system with headtracking reviewed here and described in detail in [47] is effective, it requires a significant level of hardware to be introduced into the vehicle, particularly if control is required at each seat location. Additionally, the DSP requirements are relatively high when using a large number of reference signals to provide a sufficient multiple-coherence across the control bandwidth.

4 Active Control of Reproduced Sound

In addition to the use of active sound control to attenuate, or manipulate sources of noise in the automotive environment, the physical basis of active control and the associated technologies have also been used to enhance the functionality of the infotainment system [19–24, 38]. This section will briefly review a number of systems that have been implemented for the control of audio reproduction in car cabins, with the aim of enhancing the occupants' experience.

4.1 Personal Sound Zones

The generation of independent listening zones in the car cabin without the use of headphones is a particular area of in-car sound reproduction that relies heavily on the principles of active sound control. In essence, arrays of loudspeakers are used to focus audio content to particular locations in the vehicle, whilst cancelling that content in other locations and thus allowing the occupants of the car cabin to receive personalised audio content. For example, the driver may require navigation instructions, whilst a rear-seat occupant may wish to make a private phone call. While the generation of personal sound zones can to a certain extent be achieved using directional sound sources [48], it is usually necessary to use some form of active control to achieve control at lower frequencies [22, 49, 50].

There have been a number of in-car personal sound zone systems developed in both academic and commercial research settings, with early practical demonstrations in the open-literature being reported in around 2012–13 [22, 49]. In common with the limits of ANC in the car-cabin discussed earlier in this chapter, personal audio systems using the standard car audio loudspeaker have been shown to be limited to achieving zonal separation up to around 300 Hz [22, 49]. To overcome this limitation, it has been necessary to develop methods of controlling the sound field at higher frequencies, in a similar way to high frequency local ANC. However, in the context of audio reproduction, the constraints on phase-matching a primary sound field do not exist and, therefore, there is slightly more flexibility in the active sound generation methodology at higher frequencies. As a result, a variety of methods of generating personal sound zones at higher frequencies have been proposed. For example, in

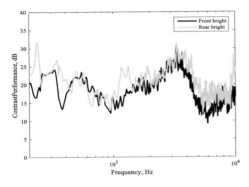

Fig. 8 The acoustic contrast achieved between the front and rear listening zones in a car cabin (left) using an array of directional, ceiling mounted loudspeakers (right) [50]

[51], headrest mounted loudspeakers were used to generate personal sound zones between adjacent seats, although the performance of the active system was limited to frequencies below around 2 kHz and at higher frequencies relied on the passive separation. To enhance the passive separation, in [22] phase-shift loudspeakers with a hypercardioid-like directivity were utilised to enhance the high frequency zonal separation between front and rear seat listening zones. This system was able to achieve more than 20 dB of separation up to 10 kHz between the front and rear zones when generating a listening zone in the front seats, but the performance was limited at higher-frequencies when generating a rear listening zone. To overcome this limitation, an alternative approach using directional ceiling mounted loudspeakers was explored in [50]. This system is shown in Fig. 8 along with the acoustic contrast, which defines the level of separation between the two listening zones. From these results it can be seen that consistent performance is achieved for both the front and rear listening zones. Whilst the levels of separation between the two listening zones reported in [50] is significant, it has been demonstrated that the subjective requirements on generating personal sound zones are often more demanding [52].

In addition to meeting the subjective requirements of personal sound zone systems, the commercial implementation of such systems requires significant levels of development beyond that often considered in academic feasibility studies. For example, in [53], it is demonstrated that practical variations in temperature in a car cabin significantly degrade the performance of a personal sound zone system. Nevertheless, a number of car manufacturers are advertising their capabilities in the generation of sound zones and forecast mass production systems by 2020 [54].

4.2 Spatial Audio

In addition to the realisation of separate listening zones in the car cabin, there is also significant interest in enhancing the infotainment system by offering spatial audio reproduction [19–21]. This could be used to improve the entertainment facilities offered to passengers, or provide an enhanced interface for the driver by, for example, using spatial audio to generate spatially localised auditory information such as navigational instructions. This could be achieved using headrest mounted loudspeakers, as in the case of local ANC or sound zone generation [55], but this may result in the issue of front-back confusion and externalisation of the sound field. To potentially overcome these issues, spatial audio reproduction can be achieved using distributed arrays of loudspeakers [20, 21], or linear arrays of loudspeakers [56], as shown in Fig. 9.

The system shown in Fig. 9 was developed to provide transaural audio reproduction for two listeners located side-by-side, as represented by the two dummy heads visible in the right-hand image of Fig. 9. This system relies on Cross Talk Cancellation (CTC) technology to generate binaural audio at the users' ears. For example, Fig. 10 shows the simulated sound field generated by the loudspeaker array in a free-field environment at two frequencies when the array is driven to generate a signal at the right ear of one user. To provide binaural audio content to both users would require this process to be repeated for each of the four ears. The practical system was

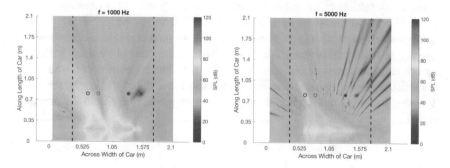

Fig. 9 A linear array of 27 loudspeakers used to generated personal spatial audio located in a car (left) and in the institute of sound and vibration research (ISVR) anechoic chamber [56]

Fig. 10 Predicated free-field pressures produced by the linear loudspeaker array shown in Fig. 9 when optimised to generate a signal at the right ear of one user [56]

tested in both an anechoic chamber and in a car cabin and was shown to achieve sufficient levels of CTC for high quality spatial audio reproduction in both environments at frequencies above around 500 Hz. The low frequency performance was limited by the loudspeaker driver selection and practical limitations on the length of the array. It was also observed that strong reflections in the car cabin environment limited the subjective performance and it was suggested that the generation of different audio programmes for the two listeners was not yet of high quality [56].

4.3 Sound Generation for Electric and Hybrid Vehicles

In addition to the use of active sound control to generate controlled sound fields within the car cabin, this technology has also been used to generate directional warning sounds, or Acoustic Vehicle Alerting Systems (AVAS) for hybrid and electric vehicles [57–59]. This technology aims to focus the warning sound required by hybrid and electric vehicles at low speeds towards the road users that require the sound, rather than unnecessarily increasing noise pollution. Various approaches to this sound control problem have been considered, including the use of low-cost directional sources [57], loudspeaker arrays [58] and arrays of structural actuators that can be readily integrated into the vehicle body [59]. Future work in this area is also on-going, with outstanding questions on how such directional warning sound systems will be viewed under the warning sound regulations.

5 Summary and Outlook

This chapter has provided a review of active sound control technology as utilised in the automotive environment. Although this technology has a well-established grounding, it has seen recent expansion due to its ability to adapt to the requirements of emerging vehicle powertrains, as well as providing additional performance and functionality in infotainment systems. This is forecast to become increasingly important as alternative mechanisms of car ownership become common and autonomous vehicles become mainstream. Although active sound control systems with high performance have been demonstrated for the control and generation of sound fields, there still remain barriers to widespread adoption due to the cost of implementation. However, it is clear that such systems are becoming available within the premium market.

References

1. Nelson PA, Elliott SJ (1991) Active control of sound. Academic press
2. Rumsey F (2013) Sound field control. J Audio Eng Soc 61(12):1046–1050
3. Rumsey F (2012) Spatial audio. Routledge
4. Druyvesteyn WF, Garas J (1997) Personal sound. J Audio Eng Soc 45(9):685–701
5. Kuo SM, Morgan DR (1996) Active noise control systems. Wiley, New York, pp 17–199
6. Elliott SJ (2010) Active noise and vibration control in vehicles. In: Vehicle noise and vibration refinement. Woodhead Publishing, pp 235–251
7. Elliott SJ (2000) Signal processing for active control. Elsevier
8. Hansen C, Snyder S, Qiu X, Brooks L, Moreau D (2012) Active control of noise and vibration. CRC Press
9. Paul L (1936) U.S. Patent No. 2,043,416. U.S. Patent and Trademark Office, Washington, DC
10. Kuo SM, Mitra S, Gan WS (2006) Active noise control system for headphone applications. IEEE Trans Control Syst Technol 14(2):331–335
11. Lester HC, Fuller CR (1990) Active control of propeller-induced noise fields inside a flexible cylinder. AIAA J 28(8):1374–1380
12. Elliott SJ, Nelson PA, Stothers IM, Boucher CC (1990) In-flight experiments on the active control of propeller-induced cabin noise. J Sound Vib 140(2):219–238
13. Boucher CC, Elliott SJ, Baek KH (1996) Active control of helicopter rotor tones. In: Inter-noise 96 (Noise control: the next 25 years, Liverpool, 30 July–2 August 1996), pp 1179–1182
14. Elliott SJ, Stothers IM, Nelson PA, McDonald AM, Quinn DC, Saunders T (1988) The active control of engine noise inside cars. In: INTER-NOISE and NOISE-CON congress and conference proceedings, vol 1988, no 3. Institute of Noise Control Engineering, pp 987–990
15. Sutton TJ, Elliott SJ, McDonald AM, Saunders TJ (1994) Active control of road noise inside vehicles. Noise Control Eng J 42(4):137–147
16. Hasegawa S, Tabata T, Kinoshita T (1992) The development of an active noise control system for automobiles. Society Automotive Eng, Technical Paper, 922 086
17. Linden A, Fenn J (2003) Understanding Gartner's hype cycles. Strategic Analysis Report No R-20–1971. Gartner, Inc., 88
18. Schirmacher R (2010) Active noise control and active sound design-enabling factors for new powertrain technologies (No. 2010-01-1408). SAE Technical Paper
19. Smithers MJ (2007) Improved stereo imaging in automobiles. In: Audio engineering society convention 123. Audio Engineering Society
20. Bai MR, Lee CC (2010) Comparative study of design and implementation strategies of automotive virtual surround audio systems. J Audio Eng Soc 58(3):141–159
21. Brix S, Sladeczek C, Franck A, Zhykhar A, Clausen C, Gleim P (2012) Wave field synthesis based concept car for high-quality automotive sound. In: Audio engineering society conference: 48th international conference: automotive audio. Audio Engineering Society
22. Cheer J, Elliott SJ, Gálvez MFS (2013) Design and implementation of a car cabin personal audio system. J Audio Eng Soc 61(6):412–424
23. Read J, Wehmeyer A (2017) Self-driving cars: a renaissance for spatial sound? In: Audio engineering society conference: 2017 AES international conference on automotive audio. Audio Engineering Society
24. https://www.dirac.com/dirac-blog/future-of-automotive-audio-blog. Accessed 10 Feb 2020
25. Elliott S, Stothers IANM, Nelson P (1987) A multiple error LMS algorithm and its application to the active control of sound and vibration. IEEE Trans Acoust Speech Signal Process 35(10):1423–1434
26. Rafaely B, Elliott SJ, Garcia-Bonito J (1999) Broadband performance of an active headrest. J Acoustic Soc Am 106(2):787–793
27. Pawelczyk M (2004) Adaptive noise control algorithms for active headrest system. Control Eng Pract 12(9):1101–1112
28. Garcia-Bonito J, Elliott SJ, Boucher CC (1997) Generation of zones of quiet using a virtual microphone arrangement. J Acoust Soc Am 101(6):3498–3516

29. Elliott SJ, Cheer J (2015) Modelling local active sound control with remote sensors in spatially random pressure field. J Acoust Soc Am 137(4)
30. Joseph P, Elliott SJ, Nelson PA (1994) Near field zones of quiet. J Sound Vib 172:605–627
31. Moreau D, Cazzolato B, Zander A (2008) Active noise control at a moving virtual sensor in three-dimensions. Acoust Australia 36(3):93–96
32. Jung W, Elliott SJ, Cheer J (2017) Combining the remote microphone technique with head-tracking for local active sound control. J Acoust Soc Am 142(1):298–307
33. Elliott SJ, Jung W, Cheer J (2018) Head tracking extends local active control of broadband sound to higher frequencies. Sci Rep 8(1):1–7
34. Han R, Wu M, Gong C, Jia S, Han T, Sun H, Yang J (2019) Combination of robust algorithm and head-tracking for a feedforward active headrest. Appl Sci 9(9):1760
35. Xiao T, Qiu X, Halkon B (2020) Ultra-broadband active noise cancellation at the ears via optical microphones. Sci Rep
36. https://www.jpmorgan.com/global/research/electric-vehicles. Accessed 14 Feb 2020
37. Chang KJ, de Melo Filho R, Geraldo N, Van Belle L, Claeys C, Desmet W (2019) A study on the application of locally resonant acoustic metamaterial for reducing a vehicle's engine noise. In: INTER-NOISE and NOISE-CON congress and conference proceedings, vol 259, no 9. Institute of Noise Control Engineering, pp 102–113
38. Cheer J (2012) Active control of the acoustic environment in an automobile cabin. Doctoral dissertation, University of Southampton
39. Kuo SM (1995) Multiple-channel adaptive noise equalizers. In: Conference record of the twenty-ninth Asilomar conference on signals, systems and computers, vol 2. IEEE, pp 1250–1254
40. Rees LE, Elliott SJ (2006) Adaptive algorithms for active sound-profiling. IEEE Trans Audio Speech Lang Process 14(2):711–719
41. Patel V, Cheer J, George NV (2017) Modified phase-scheduled-command FxLMS algorithm for active sound profiling. IEEE/ACM Trans Audio Speech Lang Process 25(9):1799–1808
42. Kobayashi Y, Inoue T, Sano H, Takahashi A, Sakamoto K (2008) Active sound control in automobiles. In: INTER-NOISE and NOISE-CON congress and conference proceedings, vol 2008, no 4. Institute of Noise Control Engineering, pp 5001–5009
43. Bernhard RJ (1995) Road noise inside automobiles. Proc Active 95
44. Oh SH, Kim HS, Park Y (2002) Active control of road booming noise in automotive interiors. J Acoust Soc Am 111(1):180–188
45. Bodden M, Belschner T (2014) Comprehensive automotive active sound design-Part 1: electric and combustion vehicles. In: INTER-NOISE and NOISE-CON congress and conference proceedings, vol 249, no 4. Institute of Noise Control Engineering, pp 3214–3219
46. Cheer J, Elliott SJ (2015) Multichannel control systems for the attenuation of interior road noise in vehicles. Mech Syst Signal Process 60:753–769
47. Jung W, Elliott SJ, Cheer J (2019) Local active control of road noise inside a vehicle. Mech Syst Signal Process 121:144–157
48. Thigpen FB (2008) U.S. Patent No. 7,343,020. U.S. Patent and Trademark Office, Washington, DC
49. Berthilsson S, Barkefors A, Brännmark LJ, Sternad M (2012) Acoustical zone reproduction for car interiors using a MIMO MSE framework. In: Audio engineering society conference: 48th international conference: automotive audio. Audio Engineering Society
50. Liao X, Cheer J, Elliott S, Zheng S (2017) Design of a loudspeaker array for personal audio in a car cabin. J Audio Eng Soc 65(3):226–238
51. Elliott SJ, Jones M (2006) An active headrest for personal audio. J Acoust Soc Am 119(5):2702–2709
52. Francombe J, Mason RD, Dewhirst M, Bech S (2012) Determining the threshold of acceptability for an interfering audio programme. Audio Engineering Society Preprint, 8639
53. Olsen M, Møller MB (2017) Sound zones: on the effect of ambient temperature variations in feed-forward systems. In: Audio engineering society convention, vol 142. Audio Engineering Society

54. https://www.hyundai.news/eu/brand/hyundai-motor-company-showcases-next-generation-separated-sound-zone-technology/. Accessed 21 Feb 2020
55. Elliott S, House C, Cheer J, Simon-Galvez M (2016) Cross-talk cancellation for headrest sound reproduction. In: Audio engineering society conference: 2016 AES international conference on sound field control. Audio Engineering Society
56. House C, Dennison S, Morgan D, Rushton N, White G, Cheer J, Elliott S (2017) Personal spatial audio in cars: development of a loudspeaker array for multi-listener transaural reproduction in a vehicle. In: Reproduced sound: sound quality by design
57. Cheer J, Birchall T, Clark P, Moran J, Fazi F (2013) Design and implementation of a directive electric car warning sound. In: Proceedings of the institute of acoustics, vol 35. Institute of Acoustics
58. van der Rots R, Berkhoff A (2015) Directional loudspeaker arrays for acoustic warning systems with minimised noise pollution. Appl Acoust 89:345–354
59. Kournoutos N, Cheer J (2020) A system for controlling the directivity of sound radiated from a structure. J Acoust Soc Am 147(1):231–241

Motivating Drivers to Drive Energy Efficient

Sandra Trösterer and Peter Mörtl

Abstract Increasing the driving range of electric vehicles has been an important research endeavour to increase the wide acceptability of electric vehicles as means for more sustainable mobility. One specific stream of research is concerned with helping to improve the driver's individual driving style for increased, but also more predictable, vehicle range. Thereby, traditional approaches often utilize displays such as eco-driving indicators or route planning capabilities, but leave a driver's motivation toward more sustainable driving often untouched. In this study, we complement this existing research by investigating on-board displays to increase drivers' motivation toward efficient driving to achieve more long-term sustainable behavioral changes. Within the European project DOMUS, we developed a tablet application that aims at motivating drivers to drive more energy efficient by using a gamification approach. Drivers are challenged to drive energy efficient during a drive, thereby competing with other drivers. In a between-subjects driving simulator study, we compared this approach with other ways to enhance driving efficiency. We compared four different conditions: (1) Drivers were asked to drive energy efficient without further information, (2) they received a training for energy efficient driving in form of an instructional video before the drive, (3) they were challenged by the application to drive energy efficient, or (4) they were provided both – training and challenges. We found that energy consumption was mostly reduced with the training video (25.04%), followed by training and use of the application (23.79%), use of the application (18.85%), and no training (15.12%).

S. Trösterer (✉) · P. Mörtl (✉)
VIRTUAL VEHICLE Research GmbH, Inffeldgasse 21a, 8010 Graz, Austria
e-mail: sandra.troesterer@v2c2.at

P. Mörtl
e-mail: peter.moertl@v2c2.at

1 Introduction

The driving range of an electric vehicle is influenced by several factors that include vehicle characteristics (such as weight, shape, and battery type), environmental factors (such as outside temperature, inclination, the kind of road, or traffic density), and active systems and energy consumers inside the vehicle (e.g., radio, air-conditioning system). In addition, the driving behaviour of the driver impacts the driving range. There exist several behavioural rules for how to save energy while driving an electric car. For example, regenerative braking (i.e., braking by releasing the gas pedal instead of using the brake pedal) may not only save energy, but can even produce energy (recuperation). Furthermore, optimal acceleration, foresight driving, smooth driving, and adequate route planning can have a positive influence on the driving range.

However, a crucial question is how we can encourage drivers to drive (more) energy efficient. There are quite a lot of existing approaches aiming at nudging eco-friendly driving behavior. Most often, drivers are provided with immediate or delayed feedback or certain actions are recommended (e.g., [3, 6, 7, 13]). Thereby, feedback may be provided in different ways, i.e., visual, haptic, auditory or subliminal feedback (e.g., [5, 12, 16]). Other approaches are more focusing on using training or tutorials to convince drivers to drive more eco-friendly or energy efficient (e.g., [1, 2]), or drivers were just told to drive fuel efficient (e.g., [9, 10]).

In our research, we focus more on what people drives to change behavior. How can we create intrinsic motivation (i.e., doing something because it's inherently interesting for a person [14, 18]) to drive energy efficient?

Following a gamification approach [11, 17], we developed an application, which challenges drivers to drive energy efficient while providing them the possibility to compare their performance with others. Figure 1 provides an overview of the general flow of the application. The application aims at increasing the drivers' intrinsic motivation by allowing drivers to explore on their own how they could reduce energy consumption and increase their driving range. At the same time, they are in competition with others, which may nudge them to further improve their driving style [4, 8]. The app is furthermore designed to keep visual distraction at a minimum while the driver is in motion.

We compared this approach with a training provided in form of an instructional video, provided before the drive. In the video, the benefits of energy efficient driving are outlined (to target drivers' intrinsic motivation) and three rules, how a driver can drive energy efficient are described (regenerative braking, anticipatory, and smooth driving). In a further condition, the instructional video and the application were combined, since being aware of behavioural ground rules most likely leads to earlier success in the challenges, which may be additionally motivating.

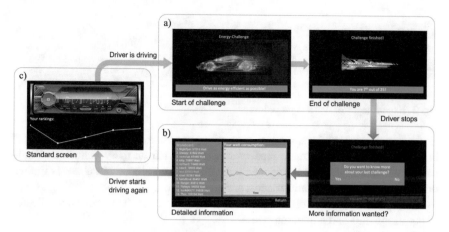

Fig. 1 General flow of the application: **a** While the driver is driving, s/he is challenged to drive as energy efficient as possible. At the end of a challenge, a short message is displayed indicating the driver's ranking in comparison to other application users. **b** If the driver stops (e.g., at a red traffic light), more information about the respective challenge becomes available. **c** In the standard screen, the scores for completed challenges are displayed, indicating the driver's progress

We had the following main research questions:

RQ1: Can energy consumption be decreased with the different methods and feedback?

RQ2: How do the different methods affect the relative energy consumption? In comparison, which method leads to the highest energy reduction?

2 Method

The study was conducted in a static driving simulator (VI-Grade) with a 180° semicircular screen. The dashboard of the car-mockup was modified to show instantaneous watt consumption like it's currently done in electric cars (eco-speedometer, see Fig. 3). Driving data was captured and processed in real-time for both, the eco-speedometer and the application.

2.1 Experimental Design

The study was realized as between-subjects design (see Fig. 2). The independent variable was the method used to bring drivers to drive energy efficient that was provided for each group of subjects: (1) no training, (2) training (instructional video), (3) challenges (provided by the app), and (4) training plus challenges.

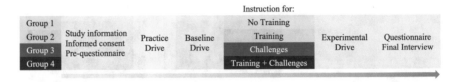

Fig. 2 Experimental design and procedure

As dependent variables, watt consumption, speed, acceleration, and number of brakes were captured as main measurements. Apart from these behavioural indicators, we also captured participants' subjective impression about changes in their driving behaviour, information needs, acceptance of energy efficient driving, as well as usefulness, fun, and information value of the application. This data was collected in questionnaires and a final interview.

2.2 Procedure

The participant was welcomed and informed about the general outline of the study. S/he then signed an informed consent and was asked to fill in a questionnaire regarding demographic data, pre-experience with electric cars and usual driving behavior. After that, the participant was asked to sit down in the driver's seat of the car mockup, to adjust the seat, buckle up, and to drive for some time in order to get used to driving in the simulator (practice drive). For the baseline drive, the participant was asked to drive as s/he would usually drive. After the drive was finished, the participant was instructed for the experimental drive according to the preassigned condition.

In addition to these instructions, each participant was told to follow the course of the road, to stay on the right lane at all times and not to overtake cars in front of them (in order to ensure that participants encounter stop and go traffic). After completing the experimental drive, participants were asked to fill in another questionnaire about eventual changes in their driving behavior, their attitude towards energy-efficient driving, and their impression about the application (in conditions 4 and 5 only). Finally, a semi-structured interview was conducted. Participants were thanked for their participation and received 20 Euro as a reward. The overall duration of one study trial was about 60–90 min depending on the respective condition.

2.3 Participants

In total, 35 subjects participated in the study. However, three participants needed to be excluded from the data analysis; two of them because they got simulator sick and could not finish the study trial and one because of technical problems during the study trial, i.e., the data of 32 participants were used for further analysis.

Table 1 Age and gender distribution in the four groups

Group	Mean age (SD)	Female	Male	Total
No training	28.50 (5.45)	4	4	8
Training	23.75 (3.57)	3	5	8
Challenges	25.00 (3.21)	4	4	8
Training + challenges	25.25 (6.36)	3	5	8

The number of participants per group was 8. Overall, the mean age was 25.6 years (SD = 4.92), with the youngest participant being 19, and the oldest 40 years old. Fourteen participants were female, eighteen male. Table 1 provides an overview of the age and gender distribution in the four groups.

All participants possessed a driving license. The average mileage was 6,771 km/year (SD = 7,777.55) with a minimum of 50 and a maximum of 25,000 km/year. Regarding the frequency of driving, 31.3% ($n = 10$) of the participants stated to drive several times a year, 43.8% ($n = 14$) several times a month, 21.9% ($n = 7$) several times a week, and 3.1% ($n = 1$) daily.

About a fifth of participants ($n = 6$; 18.8%) indicated to have driven with an electric car once before, and 68.8% ($n = 22$) stated to have went with someone in an electric car before. More than a half of the participants ($n = 18$; 56.3%) had informed themselves about electric cars at least once. We also asked participants, whether they could imagine their next car being an electric car. 75% ($n = 24$) of the participants answered this question with yes.

In order to better characterize the study participants in terms of their personal driving values, we asked them to fill in the Personal Driving Values questionnaire [15]. In the questionnaire, participants had to rate how often they drive on a highway for a stated reason on a 7-point scale (1 = never, 7 = (M=5.6) always). Our participant (M=5.2) sample was characterized by drivers, who find safe driving and avoiding fines as most (M=4.7) important, as indicated by the respective high mean ratings (M=3.1). Also, sustainable (M=3.8) driving and efficient driving is of importance, while driving fun and relaxed driving received lower ratings.

2.4 Materials and Technical Setup

As outlined, the study was conducted in our driving simulator lab. For the study, an experimental driving track was setup, which contained several speed limit and end-of-speed-limit signs, traffic lights, and three sections with stop and go traffic. The track started in a country environment (with a common speed limit of 100 km/h) and ended in a city (with a common speed limit of 50 km/h). For the stop and go sections, a virtual car was placed on the driving lane and was programmed to execute

Fig. 3 Top: Tablet with application mounted in the car mockup of the driving simulator; Bottom: Eco-speedometer (left) integrated in the dashboard

stop and go behavior once the driver passed an invisible trigger point beforehand. After a predefined interval, the virtual car turned right in a side street, allowing the driver to pass.

Along the track, further invisible trigger points were set in order to control the application. In total, five challenges were triggered, two of them requiring the driver to solely adapt to different speed limits along the track and three of them additionally requiring the driver to react to stop and go traffic.

The application was presented on a tablet with a 10.1 inch. screen running Android. The tablet was mounted in the center console of the car mockup (see Fig. 3). For the purpose of the study, the score range and number of competitors was preset for each challenge. That is, the driver's performance in a challenge (determined in real-time) was ranked in comparison to the fictive scores of 24 fictive competitors in the respective challenge. This ensured that each participant had the same initial position and same database to compare with.

For the conditions with training, an instructional video was prepared, outlining the advantages of energy efficient driving and describing three rules how a driver can drive energy efficient ((1) Use the brake pedal as little as possible, (2) Drive as anticipatory as possible, (3) Drive as smooth as possible). The video was based on a PowerPoint presentation and included audio, i.e., a speaker explained the contents verbally, accompanied by some unobtrusive background music. The video was iterated a few times based on internal feedback about understandability, pace, and sound quality. The final video lasted about three minutes.

3 Results

3.1 Driving Performance

With our first research question, we wanted to find out whether the different conditions led to a more energy efficient driving behavior compared to the baseline drive, where participants were asked to drive as they would normally do. For a fair comparison of the data, we summarized driving performance data gained during three challenges driven in the baseline drive and the corresponding three challenges in the experimental drive. Figure 4 provides an overview of the total watt consumption in watt seconds (Ws) for the baseline and the experimental drive for each condition. It is apparent that there is a decrease in total watt consumption for each condition, indicating that the participants drove more energy efficient after they were instructed accordingly. One-sided paired t-tests showed that these differences are significant for each condition (No training, $p < 0.01$; training, $p < 0.001$; challenges, $p < 0.05$; training + challenges, $p < 0.05$).

We further looked at the braking behavior of the participants during the baseline and the experimental drive. We found a large decrease in number of brakes in each condition. The differences are highly significant in each condition (all $p < 0.01$).

Regarding the speeding behavior of the participants, we were interested whether there is a change in mean speed and the variance of speed. For mean speed, we found

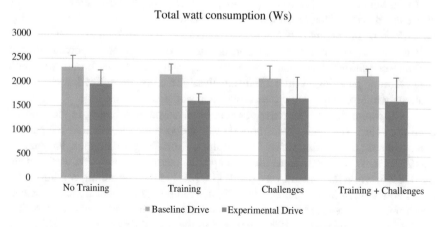

Fig. 4 Total watt consumption in the baseline and the experimental drive

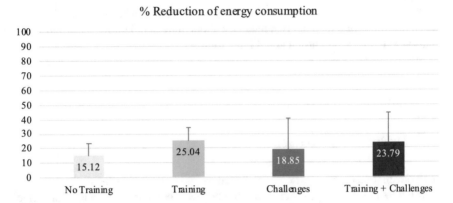

Fig. 5 Reduction of energy consumption in % for each condition

a significant decrease in speed for the experimental drives compared to the baseline drive (no training, $p < 0.01$; training, $p < 0.001$; challenges, $p < 0.05$; challenges + training, $p < 0.01$). Additionally, we also found a decrease in the mean standard deviation in speed for each condition (no training, $p < 0.001$; training, $p < 0.001$; challenges + training, $p < 0.001$), except the challenges condition (n.s.).

As a further indicator of energy-efficient driving behavior, we finally looked at the mean standard deviation of acceleration. Again, we found a highly significant decrease in the variance of the acceleration behavior (all $p < 0.01$).

In summary, these results clearly depict that there was a change of the driving behavior after the participants were instructed according to the condition. Total watt consumption significantly decreased, participants were braking less often, reduced their speed, and the variance in speed and acceleration became lower during the experimental drive.

To answer our second research question, i.e., how strongly the different conditions affect the driving behavior in comparison, we relativized the data by computing the difference in percent between the baseline and the experimental drive. Figure 5 provides an overview of the reduction of energy consumption in percent for the different conditions. Note that due to the small sample size per group, we could not determine meaningful inferential statistics. Therefore, the following results are cautiously interpreted based on the descriptive statistics.

It is apparent that the decrease of energy-consumption is largest in the training condition. Here, the energy-consumption was reduced by 25.04% after the participants had seen the training video. In the training + challenges condition, the amount was 23.79%. When participants used the app, the energy consumption was reduced by 18.85%, while the amount was lowest (15.12%) in the no training condition.

There are three major findings here. First, even without further training the watt consumption was reduced. Second, with proper training, the amount of energy saving can be further increased by 10% compared to the no training condition. Third, the amount of reduction in energy consumption is not as high as expected for the chal-

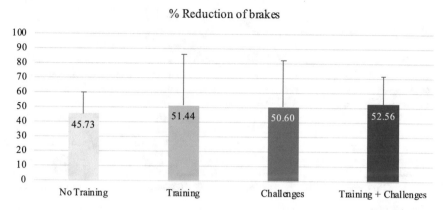

Fig. 6 Reduction of number of brakes in % for each condition

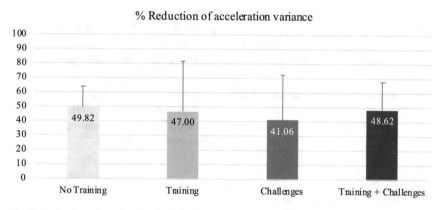

Fig. 7 Reduction of acceleration variance in % for each condition

lenges and training + challenges conditions. As depicted in Fig. 5, it seems that for those conditions also the variance in the behavior was larger (indicated by a higher standard deviation) compared to the no training and training condition. It seems that the gamification approach somehow adds another aspect, to which some people react well, while others do less.

We also looked at the reduction of brakes in percent for each condition (see Fig. 6). Here, we found a very strong change in behavior. The number of brakes was reduced by over 50% in the training, challenges, and challenges + training condition. While the amount is comparatively equal for those conditions, it was a bit lower for the no training condition with 45.73%.

With regard to the mean speed (see Fig. 8), we found that speed was reduced the most in the training + challenges condition (by 13.99%) and the no training condition (by 10.07%). For the training (5.54%) and challenges condition (7.95%), we could find less reduction. Almost the same applies for the reduction of speed variance.

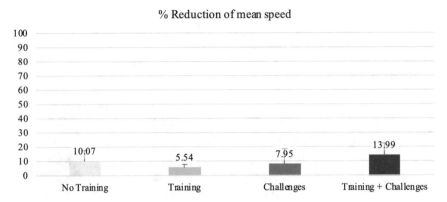

Fig. 8 Reduction of mean speed in % for each condition

Finally, we also found that the acceleration variance was strongly reduced (see Fig. 7). It was mostly reduced in the no training, training, and training + challenges condition (greater or equal 47% each), and was a bit lower in the challenges condition (41.06%).

In summary, we found that energy consumption was mostly reduced in the training condition, followed by challenges + training, challenges, and no training. A further look at the driving performance data shows that particularly the number of brakes and acceleration variance were strongly reduced in each condition.

3.2 Questionnaire and Interviews

After the experimental drive was finished, we asked participants several questions about changes in their driving behavior, information needs, and attitude towards energy-efficient driving. Regarding the question, what participants did change about their driving behavior in comparison to their normal driving behavior, most partici-pants' statements were concerned with foresighted driving, different braking behav-ior, and different acceleration. Some of them were aware from the outset, that braking should be avoided and that it is better to coast and drive anticipatory.

Regarding the question, what the biggest difference to their normal driving behav-ior were, participants mostly named changes in acceleration (less, slower), braking behavior (less, less abrupt, coasting), speeding behavior (slower), and anticipatory driving. About half of the participants sensed energy-efficient driving as restrictive ($n = 15$). "I prefer to accelerate faster". "It would be a restriction, because other drivers would have a problem with it". However, some participants ($n = 10$) consid-ered energy-efficient driving as comfortable and less stressful, while others ($n = 10$) experienced it as fun. "It was fun, because it was completely new". "It was fun in the simulator, because I didn't have stress because nobody was behind me". Although these were only singular statements, it needs to be pointed out that how other drivers may react to an energy-efficient driving style seems to be an important point.

Regarding the question, whether participants would have wished for further information about how to drive energy efficient, it was stated most often in the challenges group ($n = 11$) that further information would have been helpful. Five participants would have liked more general information "how to drive better", proper feedback in the app "What am I doing wrong in the situation?", when to accelerate, and how to brake. The information need in the no information group was less high ($n = 4$). Two participants stated that they would have liked general information how to drive energy efficient and two wanted more information how their driving style and the eco-speedometer relate.

Finally, we also asked participants how it was for them to try out energy efficient driving in the simulator and whether they got an adequate impression. More than a half of the participants stated that they got an adequate impression ($n = 18$; 56.3%). They considered the simulation as realistic ($n = 7$) and pointed out that the eco-speedometer was informative ($n = 2$), that they learned something about energy efficient driving ($n = 2$), or although the driving experience was different, they still experienced the energy efficient driving ($n = 2$).

On the other hand, 10 participants criticized the simulation primarily for being unrealistic and not providing adequate feedback about acceleration and braking. It was also mentioned that speed signs and traffic lights usually would be visible from a larger distance in reality and could be anticipated better, or that there weren't any other traffic participants in the simulation reacting to their driving style ("In reality, other cars might start to honk if I drive slower"). Some statements, however, also rather targeted the experimental track layout itself ("It's unrealistic to have so many changes of speed in a row", "It were always the same cars"). Five participants thought that the eco-speedometer was not accurate enough or that they only could rely on the eco-speedometer, in order to get feedback about their driving. Three participants also mentioned that they experienced a certain dizziness and light nausea.

In the challenges and training + challenges group, we additionally asked participants some questions about the application. First, we wanted to know how the participants generally did get on with the challenges issued by the app. Here, about half of the statements were positive, the other half negative. Five participants stated that they did get on well with the challenges and that the challenges were easy, four mentioned that they paid more attention to their driving style, and three participants stated that the ranking was interesting and pushed them to drive more energy efficient.

We further asked the participants what they liked/disliked about the app. Here, the presentation, design and the graph were mentioned positively most often ($n = 9$). "The graph was good". "Nice design". Also, the using the app was considered as easy ($n = 2$), that the app reminds one to drive energy efficient ($n = 1$), or only allows for further interaction, when one is waiting at the traffic light ($n = 1$). Regarding the question, what participants disliked about the app, the lack of feedback about the own driving behavior and lack of information how to improve it was mentioned most often ($n = 9$). "I felt the ranking was arbitrary. It did not allow to see my own performance improvement". "The app was based on social comparison. I would have preferred to get feedback about my performance. Have I myself managed to drive better?" "There was no information how to improve". Other statements were related to the graph ("The graph wasn't meaningful, because I could not see when what happened"), the visualizations, and the sound.

Regarding the question, whether participants considered the app and the challenges as useful to improve the energy efficiency of their driving style, we could find a difference between the challenges and the training + challenges group. In the former group, half of the participants ($n = 4$) found the app useful, in the latter group almost all participants ($n = 7$). Overall, it was mentioned most often, that the app is useful because it motivates ($n = 7$). "The app has a stimulative nature". "Competition stimulates". "You have an aim". It was also mentioned that one pays more attention to the driving style because of the app, or that the app provides a training, which may change driving behavior on the long run. Participants, who did not find the app useful, mostly criticized the lack of feedback and information how to improve. "I know that I do not drive well, but I don't know what I am doing wrong. I don't get constructive feedback". "I didn't get individual feedback, where I could have driven better". Other statements did not see an essential advantage over the eco-speedometer, it was doubted that it could work in reality, or that generally the challenges approach was not working for the person.

4 Discussion

4.1 Impact of the Conditions on Energy Efficient Driving Behavior

With regard to our first question, whether an energy efficient driving style can be successfully induced with the different trainings and feedback, we found that there is a significant decrease in watt consumption in each condition when comparing the baseline and the respective experimental drive. Furthermore, there is a clear reduction of number of brakes and a clear reduction of variance in acceleration observable in each condition. Also, mean speed and variance of speed was reduced in the experimental drives. In summary, we could find a considerable change in driving behavior in each condition, i.e., an energy efficient driving style could be successfully induced in each condition.

Regarding the second research question, which condition leads to the most energy efficient driving behavior, we found that total watt consumption was reduced by 25.04% in the training condition. The amount was 23.79% in the training + challenges condition, 18.85% in the challenges condition, and 15.12% in the no training condition. As outlined in the results section, we could not determine significant differences between the groups due to the sample size—therefore the results need to be interpreted with caution. However, note that the subjective impressions gained in the questionnaire and final interview support the findings and tendencies in the driving data.

Our results indicate that the training condition was most successful to reduce overall energy consumption. At the same time, it is noteworthy that total watt consumption was also reduced in the no training condition. This is also in line with [10], who found that telling participants to drive energy efficient already has an impact. However, in comparison to the no training condition, the amount of energy saving could be further increased by 10% if participants received training. It should be noted at this point that the training was straightforward and only provided three ground rules how to drive energy efficient. Our results show that even with a small training, the impact can be quite high. However, we are also aware that the participants had those rules freshly in mind when starting the experimental drive, while in reality such behavioral rules may be forgotten after some time.

Our results also show that the conditions with the challenges did not work as well as expected. While we could achieve almost the same amount in watt reduction for the training + challenges condition compared to the training condition, the challenges condition had a weaker impact. Also, we found a considerably higher variance of the data in the challenges conditions, indicating that the participants behaved more diverse in those conditions. We believe that the gamification approach triggered participants to drive energy efficient, however, participants may have applied more or less successful strategies to do so.

When comparing the conditions with regard to the different driving performance parameters, we find slight differences between the conditions, which in combination resulted in the overall differences. It is noteworthy, that particularly the number of brakes and variance in acceleration was strongly reduced in all conditions.

4.2 Strengths and Weaknesses of the Application

In general, participants indicated that they tried to master the challenges provided by the app as good as possible. However, one crucial point about the app seems to be that it does not provide enough information how to drive energy efficient. This goes hand in hand with the results based on the driving behavior, where we could find that the challenges condition only slightly outperformed the no training condition. Indeed, the app only provides the ranking and the graphical representation how watt consumption evolved over time for the respective challenge, but no feedback how to drive energy efficient or how to improve. While this was originally intended to make the game more interesting and challenging, our study showed that this approach is problematic when the app is used without additional training or information.

While the graph about watt consumption over time and the presentation was generally experienced positive, the lack of information and feedback how to drive better was mostly disliked. We also found that the ranking was problematic for several reasons. On the one hand, it was unclear for some participants how the ranking worked and who the other people were, on the other hand, we also found that the social comparison was not motivating or interesting for some participants. They rather wished to get information about the individual performance and whether it increased/decreased from challenge to challenge. While we found in related work that social comparison can be a strong motivator, our results show that this may not apply for everyone and the app currently supports drivers too little in challenging themselves. However, we are also aware that the artificiality of the scoreboard was obviously a factor that impacted the motivating aspect of social comparison in the study. In reality, one might also compete with friends or family, which could be more motivating.

Also, we found that the attitude towards the app in general seemed to be more positive when it was accompanied with training. We also could find split opinions about the challenges in the final interviews. Some participants found them good, while others did not find them to be challenging. However, it was also outlined that the app is generally motivating and reminds one to pay attention to the own driving style and to drive energy efficient. This is probably a crucial point, because in reality nobody will remind the driver to drive energy efficient, although we could see in the no training condition that a simple instruction can already make a difference. From that point of view, the strength of the app certainly is that it serves as such a reminder. However, the transfer of knowledge – "What do I need to do in order to drive more energy efficient" – needs to better implemented. Our results showed that only half of the participants in the challenges group found the app to be useful to drive energy efficient without this information, while almost all participants found the app useful, when they were provided with additional training.

Since the thumb rules provided in the training video showed to be quite successful, we could imagine that one approach could be to bring in these messages in the app. For example, the message when the challenge starts ("Drive as energy efficient as possible") could be replaced with more goal-oriented messages, e.g., "Try to use the brake pedal as least as possible". In this way, drivers would learn what the behavioral rules are, and which effect they can have on energy efficiency. After some time and with enough training, it then probably would be possible to use more general messages again.

4.3 Driving Energy Efficient in the Simulator Versus in Reality

Based on their experiences in the driving simulator, about half of the participants found energy efficient driving as a restriction of their own driving style, while the other half thought that it is comfortable, less stressful and is fun. At the same time, we also found that participants did not perceive energy efficient driving as a strong contradiction to their own driving style, nor was it particularly challenging for them to change their driving style. Also, it was indicated in the questionnaire that they rather could get used to an energy efficient driving style.

More than half of the participants stated that they gained an adequate impression about energy efficient driving in the simulator. However, there was also some criticism regarding the simulation and the experimental track. Particularly, it was criticized that the simulation did not provide adequate feedback for braking and acceleration and that the visibility of speed signs from a distance was limited. We are aware that this is a general limitation of simulator studies and also highly depends on the used simulator. However, it needs to be outlined that our study was setup as a comparative study, i.e., the given restrictions by the simulated environment were present in each condition and drive equally, allowing a fair comparison within the study.

One critical point, though, when it comes to driving energy efficient in real traffic, could be the reaction of other traffic participants. In the simulation, we only had upcoming traffic implemented, but no car was behind the participant. However, it was stated by some participants that this was unrealistic and would probably impact their behavior in reality, because in real traffic characteristics of an energy efficient driving style might be less accepted by other traffic participants and may cause annoyed reactions. We believe that this social factor is indeed important and needs further consideration.

Other points were concerned with the experimental track itself and the accuracy of the eco-speedometer. For the study, we needed to design the track in a way that a lot of variation in driving behavior could be induced in a short time interval, since longer drives may also increase the probability of simulator sickness. Therefore, the track had to be artificial to a certain degree in order to ensure this variability.

4.4 Limitations

Since the implementation of the technical setup took substantially longer than anticipated, we could not conduct the study with the number of participants we originally planned. However, we think that the combination of quantitative and qualitative data provided us with a comprehensive picture how the different conditions impacted energy efficient driving behavior.

As outlined, our participant sample had a rather positive attitude towards sustainable driving from the outset, and to a certain degree could rely on some previous knowledge about fuel efficient driving gained in driving school. We are aware that this probably also has a certain impact on our results, and conditions without further provided information may have profited from the sample characteristics. However, we would not expect a change of our overall results, only that the benefit of training could have probably been shown even stronger.

Since our study was conducted in the simulator with a particular experimental track and lack of other (reacting) traffic participants, we are aware that there are limitations with regard to the ecological validity of our results. However, when comparing the drives, we could see clear changes in driving behavior and tendencies in the data, which are transferable to reality, although the absolute values may differ.

Another limitation of the reported study is that it observed only short term effects of the energy increasing interventions; it did not observe the driving over a longer term such as days, weeks, or even months. While the measured effects may have been visible immediately after the efficiency increasing interventions, they may weaken or disappear as drivers find back to their habitual ways of driving.

5 Conclusions

Our results show that an energy efficient driving style could be successfully induced with all methods compared in the study. Particularly, in the conditions involving training the energy consumption was reduced most. While the application motivated participants to focus on energy efficient driving behavior, further information about how to drive energy efficient and how to improve needs to be integrated. Also, the possibility to not only compete with others but also with oneself needs to be integrated as adjustable gamification setting.

Our study showed that it is possible for drivers to drive efficiently without any difficulties, however, the challenge remains, how to convince drivers in the real world to do so. The study showed that tutoring and gamification may help, however, in the real world further factors, like the impact of the behavior and reactions of other traffic participants, and long-term effects of tutoring and gamification need further investigation.

Acknowledgments We thank Stefan Erlachner, Hassaan Islam, Nik Adzic, and Sebastian Möller for the technical realization of the study, and Nikolai Ebinger and Nino Dum for supporting the study preparation and conduction.

This project has received funding from the European Union's Horizon 2020 research and innovation programme under grant agreement No. 769902. The publication was written at the VIRTUAL VEHICLE Research GmbH in Graz and partially funded by the COMET K2—Competence Centers for Excellent Technologies Programme of the Federal Ministry for Transport, Innovation and Technology (bmvit), the Federal Ministry for Digital, Business and Enterprise (bmdw), the Austrian Research Promotion Agency (FFG), the Province of Styria and the Styrian Business Promotion Agency (SFG).

References

1. Barla P, Gilbert-Gonthier M, Lopez Castro MA, Miranda-Moreno L (2017) Eco-driving training and fuel consumption: impact, heterogeneity and sustainability. Energy Econ 62:187–194. https://doi.org/10.1016/j.eneco.2016.12.018. https://linkinghub.elsevier.com/retrieve/pii/S0140988317300051
2. Beloufa S, Cauchard F, Vedrenne J, Vailleau B, Kemeny A, Mérienne F, Boucheix JM (2019) Learning eco-driving behaviour in a driving simulator: contribution of instructional videos and interactive guidance system. Trans Res Part F Traffic Psychol Behav 61:201–216. https://doi.org/10.1016/j.trf.2017.11.010. https://linkinghub.elsevier.com/retrieve/pii/S1369847816304351

3. Dahlinger A, Wortmann F (2016) Towards the design of eco-driving feedback information systems: a literature review, p 12. https://pdfs.semanticscholar.org/ab68/9feec2b7692f3af9f05fe4cc866cc8db1b57.pdf

4. Ecker R, Holzer P, Broy V, Butz A (2011) EcoChallenge: a race for efficiency. In: Proceedings of the 13th international conference on human computer interaction with mobile devices and services—MobileHCI'11, p 91. ACM Press, Stockholm, Sweden. https://doi.org/10.1145/2037373.2037389. http://dl.acm.org/citation.cfm?doid=2037373.2037389

5. Hammerschmidt J, Tünnermann R, Hermann T (2014) EcoSonic: auditory displays supporting fuel-efficient driving. In: Proceedings of the 8th Nordic conference on human-computer interaction fun, fast, foundational—NordiCHI'14, pp 979–982. ACM Press, Helsinki, Finland. https://doi.org/10.1145/2639189.2670255. http://dl.acm.org/citation.cfm?doid=2639189.2670255

6. Hibberd D, Jamson A, Jamson S (2015) The design of an in-vehicle assistance system to support eco-driving. Trans Res Part C Emerg Technol 58:732–748. https://doi.org/10.1016/j.trc.2015.04.013. https://linkinghub.elsevier.com/retrieve/pii/S0968090X15001540

7. Kircher K, Fors C, Ahlstrom C (2014) Continuous versus intermittent presentation of visual eco-driving advice. Trans Res Part F Traffic Psychol Behav 24:27–38. https://doi.org/10.1016/j.trf.2014.02.007. https://linkinghub.elsevier.com/retrieve/pii/S1369847814000217

8. McConky K, Chen RB, Gavi GR (2018) A comparison of motivational and informational contexts for improving eco-driving performance. Trans Res Part F Traffic Psychol Behav 52:62–74. https://doi.org/10.1016/j.trf.2017.11.013. https://linkinghub.elsevier.com/retrieve/pii/S1369847816301139

9. Pampel SM, Jamson SL, Hibberd D, Barnard Y (2017) The activation of eco-driving mental models: can text messages prime drivers to use their existing knowledge and skills? Cogn Technol Work 19(4):743–758. https://doi.org/10.1007/s10111-017-0441-3

10. Pampel SM, Jamson SL, Hibberd DL, Barnard Y (2018) Old habits die hard? The fragility of eco-driving mental models and why green driving behaviour is difficult to sustain. Trans Res Part F Traffic Psychol Behav 57:139–150. https://doi.org/10.1016/j.trf.2018.01.005. https://linkinghub.elsevier.com/retrieve/pii/S1369847817300414

11. Piccolo LSG, Baranauskas MCC (2012) Requirements for situated eco-feedback technology, p 10

12. Riener A (2012) Subliminal persuasion and its potential for driver behavior adaptation. IEEE Trans Intell Trans Syst 13(1):71–80. https://doi.org/10.1109/TITS.2011.2178838

13. Rolim C, Baptista P, Duarte G, Farias T, Pereira J (2016) Impacts of delayed feedback on eco-driving behavior and resulting environmental performance changes. Trans Res Part F Traffic Psychol Behav 43:366–378. https://doi.org/10.1016/j.trf.2016.09.003. https://linkinghub.elsevier.com/retrieve/pii/S1369847816302868

14. Ryan RM, Deci EL (2000) Self-determination theory and the facilitation of intrinsic motivation, social development, and well-being. Am Psychol 67

15. Shahab Q, Terken J, Eggen B (2013) Development of a questionnaire for identifying driver's personal values in driving. In: Proceedings of the 5th international conference on automotive user interfaces and interactive vehicular applications—AutomotiveUI'13, pp 202–208. ACM Press, Eindhoven, Netherlands. https://doi.org/10.1145/2516540.2516548. http://dl.acm.org/citation.cfm?doid=2516540.2516548

16. Staubach M, Kassner A, Fricke N, Schiessl C, Brockmann M, Kuck D (2012) Driver reactions on ecological driver feedback via different HMI modalities, p 13

17. Steinberger F, Schroeter R, Foth M, Johnson D (2017) Designing gamified applications that make safe driving more engaging. In: Proceedings of the 2017 CHI conference on human factors in computing systems—CHI'17, pp 2826–2839. ACM Press, Denver, Colorado, USA. https://doi.org/10.1145/3025453.3025511. http://dl.acm.org/citation.cfm?doid=3025453.3025511

18. Tulusan J, Steggers H, Staake T, Fleisch E (2012) Supporting eco-driving with eco-feedback technologies: recommendations targeted at improving corporate car drivers' intrinsic motivation to drive more sustainable

Energy Efficient and Comfortable Cabin Heating

Alois Steiner, Alexander Rauch, Jon Larrañaga, Mikel Izquierdo, Walter Ferraris, Andrea Alessandro Piovano, Tibor Gyoeroeg, Werner Huenemoerder, Damian Backes, and Marten Trenktrog

Abstract Battery electric vehicles become an increasingly interesting alternative to conventional vehicles with combustion engine as the costs decrease and the driving range improves. Although, the significant decrease of the driving range at very hot or cold ambient temperatures is still an issue that needs to be solved in order to further improve the adoption of battery electric vehicles. In this chapter a system for energy efficient cabin heating consisting of a heat pump system, a so-called smart seat and radiation heating panels is presented. By means of simulations the possible reduction of energy consumption and increase of driving range during an NEDC is evaluated.

List of Abbreviations

COP   Coefficient of Performance
CRU   Compact Refrigeration Unit
HVAC   Heating, Ventilation and Air Conditioning
NEDC   New European Driving Cycle
PMV   Predicted Mean Vote

A. Steiner (✉) · A. Rauch
Virtual Vehicle Research GmbH, Graz, Austria
e-mail: alois.steiner@v2c2.at

J. Larrañaga · M. Izquierdo
Faculty of Engineering, Mondragon Unibertsitatea, Arrasate, Spain

W. Ferraris
Centro Ricerche Fiat, Turin, Italy

A. A. Piovano
FCA Italy S.p.A, Turin, Italy

T. Gyoeroeg · W. Huenemoerder
DENSO AUTOMOTIVE Deutschland GmbH, Eching, Germany

D. Backes · M. Trenktrog
Institute for Automotive Engineering, RWTH Aachen University, Aachen, Germany

© The Author(s), under exclusive license to Springer Nature Switzerland AG 2021
A. Fuchs and B. Brandstätter (eds.), *Future Interior Concepts*,
Automotive Engineering : Simulation and Validation Methods,
https://doi.org/10.1007/978-3-030-51044-2_5

PPD   Predicted Persons Dissatisfied

1 Introduction

The power demand of air conditioning systems at very hot or cold ambient temperatures significantly decreases the driving range of battery electric vehicles at these conditions. At winter conditions the driving range may drop by 30–50% (see e.g. [1, 2]) because of the high power demand of resistance heaters and the decreasing efficiency of lithium-ion batteries. Thus, innovative solutions for the heating of the passenger compartment are needed. In the EU-funded research project "OPTEMUS" a solution with a compact refrigeration unit, that enables heat pump operation, a "smart seat" with integrated Peltier elements and radiation heating panels has been developed. This system significantly reduces the energy consumption for heating the passenger compartment and increases the driving range of electric vehicles.

2 Thermal Comfort Assessment

In order to enable a fair comparison of different heating technologies in terms of energy consumption, equal thermal comfort needs to be ensured. As different heating technologies use different heat transfer paths (thermal convection with the heat pump system, thermal conduction with the seat heating and thermal radiation with the radiation heating panels) a thermal comfort assessment method that considers all of them was required. The so-called Predicted Mean Vote (PMV) model, developed by Fanger [3], is one of the most recognized thermal comfort models and was used for the comfort assessment. Its comfort index ranges from −3 (cold) to +3 (hot) and is depending on six calculation factors, which are shown in Fig. 1.

The PMV is linked to the Predicted Persons Dissatisfied (PPD) with the following equation:

$$PPD = 100 - 95e^{-0.03353PMV^4 - 0.2179PMV^2}$$

This leads to the fact that at a PMV of +3 (hot) or −3 (cold) 99% of occupants feel thermal discomfort. A PMV equal to zero, which represents thermal neutrality, leads to 95% of occupants that feel thermal comfort and only 5% discomfort (Fig. 2).

For a comparison of different heating technologies in terms of energy consumption a PMV of zero was used as a constraint.

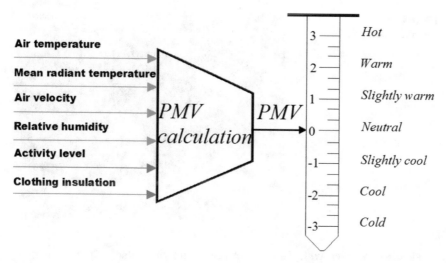

Fig. 1 Scale and calculation factors of PMV index [1]

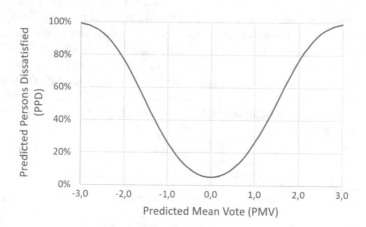

Fig. 2 Predicted persons dissatisfied (PPD) depending on predicted mean vote (PMV)

3 Heat Pump System with the Compact Refrigeration Unit

For an efficient heating of the cabin inlet air a heat pump system was used. Its core part is the compact refrigeration unit (CRU) from DENSO using a natural refrigerant. It is a so-called water-to-water system that consists of a compressor, two plate heat exchangers and an expansion valve. All components are integrated in the same housing (Fig. 3).

Inside the evaporator (chiller) heat is transferred from the coolant circuit to the refrigeration cycle and consequently cooling down the coolant. This amount of heat increased by the heat of the compressor is transferred to a second cooling circuit

Fig. 3 Compact
refrigeration unit (CRU)

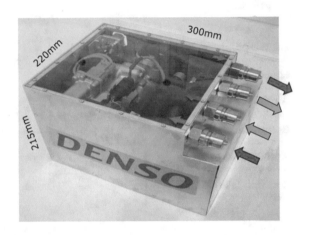

inside the condenser, with the effect of heating up the coolant. The two emerging temperature levels can be used for the thermal management in the vehicle in order to combine heat sources and heat sinks in an energy efficient way. Therefore, the system depicted in Fig. 4 was used to divert the coolant flows to various components of the electric vehicle to enable different operation modes. These operation modes are cooling, heating and dehumidification of the passenger cabin, battery heating and cooling as well as powertrain heating and cooling. Figure 4 shows an operation in heat pump mode, where the passenger cabin is heated and the ambient air is used as heat source. By using the ambient air as heat source a coefficient of performance (COP) of 1.50 could be achieved at 0 °C and a COP of 2.67 at +10 °C for cabin heating during NEDC.

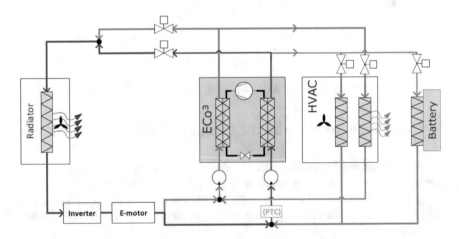

Fig. 4 Thermal management system for an electric vehicle (heat pump mode)

4 Radiation Heating Panels

Conventional heating, ventilation and air conditioning (HVAC) systems only rely on the convective heat transfer where air is conditioned before entering the passenger cabin. These systems offer a simple implementation into a vehicle and a simple possibility to use waste heat from combustion engines, but also show considerable disadvantages when assessing energy efficiency of electrified vehicles. Especially in "ambient air mode" the energy demand is high because heated air is ventilated to the environment and these enthalpy losses need to be compensated. In addition to the intake air further heat capacities like the coolant, heater core or interior components need to be heated up as well. These thermal inertias cause only little impact during long distance use cases. Considering urban electric vehicles with primarily short trips, lowering thermal inertias inside the HVAC system offers potential for improving the energy efficiency. As discussed in [4], radiation heating panels offer an approach to increase both heating efficiency and passenger comfort at the same time. While Pingel et al. [4] propose a far infrared surface heating system, the state of the art also considers infrared heating systems with higher temperatures [5]. Both heating systems are not able to replace the conventional HVAC system entirely, but to reduce its energy demand. As shown in Fig. 1, thermal comfort is depending on a variety of factors. So, if the mean radiant temperature of surrounding surfaces can be increased by radiation heating, the air temperature may be decreased. Radiative heat is transferred as electromagnetic radiation with the wavelength depending on the surface temperature of the emitter. Radiation heating systems can be considered as long, mid and short wavelength infrared, with increasingly higher temperatures of the emitter. As no consistent terminology for these heating systems exists, especially solutions using shorter wavelengths are often called infrared heaters. Other solutions, which also emit infrared radiation on lower temperature levels and thus longer wavelengths are usually called surface heating systems. Latter cover relatively large areas and emit radiation at temperatures below 70 °C in order to prevent skin injuries. Both radiation heating systems may be used for vehicle passenger cabins. In the Fiat 500e demonstrator vehicle a surface heating system was applied on different surfaces close to the driver and the co-driver. Figure 5 shows a simulation of surface temperatures of the car interior at −10 °C ambient temperature. All radiation panels were set to a constant surface temperature of 47 °C for the transient simulation of the cabin heat up, not considering the fast heat up of the panel itself.

The simulation was set up with a "virtual comfort dummy" inside the cabin, considering thermal radiation transferred from the panels to the passenger by means of view factors. With a total electric panel power of 100 W the air temperature could be lowered by 3 K for quasi-static operating conditions in heating mode while maintaining the same comfort level.

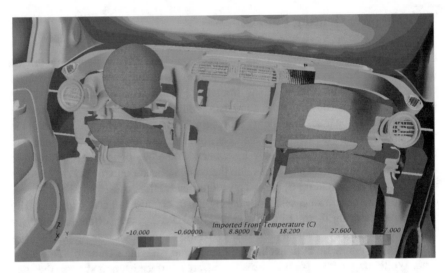

Fig. 5 Simulation of surface temperatures of the car interior at $-10\ ^\circ$C ambient temperature

5   Smart Seat

The smart seat was designed to cool or heat the passenger by direct contact depending on the climate conditions. This is achieved by the combination of thermoelectric elements (Peltier elements) and five thermally conductive layers within the smart seat specifically at the areas with direct contact with the driver/passenger. The advantage of using the seat for heating operation is the reduction of air temperature inside the cabin (vice versa in cooling operation the air temperature inside the cabin can be increased), which thus will reduce the energy consumption of the heat pump system ensuring same comfort. In order to avoid local discomfort, the heat flow should be limited to approx. 60 W/m^2 [6, 7] for the heating case. Moreover, lower body regions (e.g. thigh, foot etc.) should be warmer than the upper body regions (e.g. arm, chest etc.) [7]. Consequently the control of the smart seat has been designed accordingly. The adaptive control theory (fuzzy logic) has been applied to regulate the comfort of the user based on the PMV model. Two main aspects should be carefully considered: (i) cloth resistance estimation and (ii) heating and cooling by direct contact. ISO 9920 standard [8] establishes the method for measuring the coverage index (cloth resistance). However, this mathematical method cannot be applied to calculate it online, since the variables used to calculate it are passenger dependent (i.e. are purely stochastic variables). For this reason, an artificial intelligence algorithm has been implemented taking as input variables the ambient temperature and the relative humidity measured when the user switches on the system. Then, by measuring the temperature increase of the smart seat in during 5 s of operation, the cloth resistance is predicted. Figure 6 shows the numerical simulation of the smart seat integrated in the Fiat 500e interior after 10 min of warm up at $-10\ ^\circ$C ambient temperature with

Fig. 6 Simulation of surface temperatures of the car interior after 10 min of warm up at −10 °C ambient temperature with focus on the driver's seat

focus on driver's seat. It can be clearly observed the localized areas of the seat which are heated for the thermal comfort of the driver.

Once the smart seat was virtually validated (thermal and sitting comfort was evaluated) it was manufactured and assembled. Before integrating it in the vehicle it was tested under laboratory conditions in order to validate the concept. Figure 7 shows the surface temperature distribution of the seat in heating operation. In addition to temperatures measured in the seat, the temperature of the passenger's face was

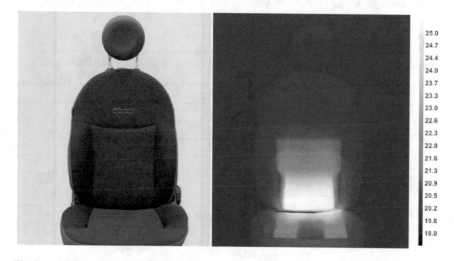

Fig. 7 Surface temperatures of the seat in heating operation

also measured. The temperature of the skin is a clear indicator to evaluate thermal sensations of humans. The metabolic system regulates the blood flow near the skin surface, the sympathetic nervous system through sweating and constriction, subcutaneous thermal structure and patterns of the facial veins [9]. The seat and the driver were heated (or cooled) as expected, proving the concept validation for the smart seat. Applying a power of 75 W to the seat enabled an air temperature reduction of 7 K in heating mode while maintaining the same comfort level for the passenger.

6 Impact on Energy Consumption and Driving Range

In order to quantify the impact on the energy consumption and driving range of the Fiat 500e simulations of a cabin heat up for the use cases in Table 1 were conducted. The simulations were run at 0 and +10 °C, which are two frequently occurring ambient temperatures during the cold season. As driving cycle the NEDC was chosen, with an average power for traction of 7 kW. During the 1180 s of the NEDC the heating power was controlled to reach the cabin target temperature at the end of the cycle. The boundary conditions for the simulated use cases are shown in Table 2, while radiation heating of 200 W represents two panels for both, driver and co-driver seat.

Figure 8 shows the simulation results for the average electric power for cabin heating during an NEDC at 0 °C ambient temperature. The use of the compact refrigeration unit as heat pump system (average COP of 1.50) reduces the average electric power from 3.51 to 2.34 kW (−33%). Using seat and panel heating in addition, which enables to lower the cabin air temperature from 22 to 12 °C further decreases the average power significantly to 0.83 kW (−76% compared to PTC heating).

At +10 °C ambient temperature the heat pump system enables a drastic reduction of the average power from 1.76 to 0.66 kW (−63%) at an average COP of 2.67

Table 1 Simulation use cases

Use case	Air heating technology	Cabin target temperature (°C)	Seat heating (75 W)	Radiation heating (200 W)
1	PTC-heater	22	No	No
2	Heat pump	22	No	No
3	Heat pump	19	No	Yes
4	Heat pump	15	Yes	No
5	Heat pump	12	Yes	Yes

Table 2 Boundary conditions for the conducted simulations

Driving cycle, duration	NEDC, 1180 s
Ambient temperature	0 and +10 °C
Air flow cabin heater	250 kg/h

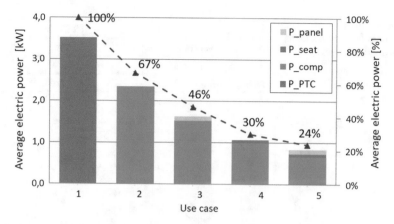

Fig. 8 Average electric power for cabin heating during an NEDC at 0 °C ambient temperature

(Fig. 9). For the use cases 3, 4 and 5 the power of the heat pump system stays the same, as the compressor is already running at minimum speed. Therefore, a further reduction of the air temperature brings only a small advantage in terms of energy consumption. Use case 4 with a cabin target air temperature of 15 °C and activated seat heating is according to the simulation results the most efficient option. The average power could be reduced from 1.76 to 0.43 kW (−76% compared to PTC heating). Use case 3 (heat pump and panel heating with a cabin temperature of 19 °C) and use case 5 (heat pump with panel and seat heating) are slightly less efficient.

In order to quantify the impact on driving range an average power for traction as well as for auxiliaries (pumps, fans etc.) was assumed (Table 3). The battery capacity of the Fiat 500e is 24.9 kWh and the average speed taken from the NEDC is 34 km/h.

Figure 10 shows the simulated driving range for the Fiat 500e for the different use cases at 0 °C ambient temperature. The driving range of 77 km using a PTC heater

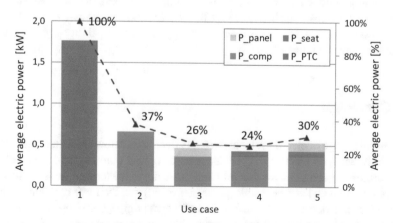

Fig. 9 Average electric power for cabin heating during an NEDC at +10 °C ambient temperature

Table 3 Boundary conditions for the calculation of the driving range

Battery capacity	24.9 kWh
Average power for traction	7 kW
Average power for auxiliaries	500 W
Average speed	34 km/h

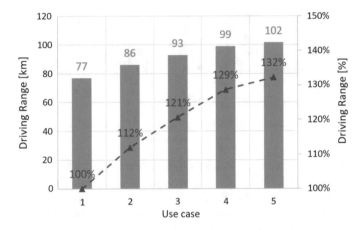

Fig. 10 Driving range for the Fiat 500e at 0 °C ambient temperature for the different use cases

can be improved up to 102 km (+32%) using a heat pump combined with seat and panel heating. To sum up, the combination of different heating technologies enables to lower the cabin air temperature from 22 to 12 °C and decreases convective heat losses.

At +10 °C ambient temperature the possible increase of driving range is lower than at 0 °C, as the power for cabin heating is generally lower (Fig. 11). Whereas PTC heating in use case 1 results in a driving range of 91 km, the heat pump system in use case 2 increases the driving range by 14% to 104 km. Use case 4 with the heat pump system and seat heating is the most energy efficient possibility at +10 °C with a resulting driving range of 107 km (+17%).

7 Conclusions

The driving range of electric vehicles strongly decreases at low ambient temperature, especially when an electric PTC heater is used. A heat pump system is one of the most promising solutions for this problem. The closer the temperatures of the heat source (ambient air) and heat sink (passenger cabin) are, the more efficient the heat pump system is operating. A combination of a heat pump system together with seat heating and radiation heating panels enables a lower air temperature inside the cabin and improves the COP of the heat pump system at same comfort level. Further,

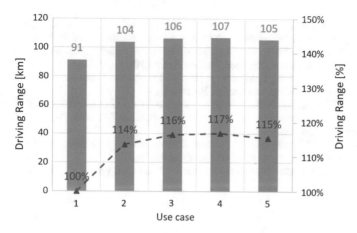

Fig. 11 Driving range for the Fiat 500e at +10 °C ambient temperature for the different use cases

the heat losses by warm air exiting the passenger cabin are reduced at lower cabin temperatures. The simulation results for a cabin heat up in an NEDC showed, that the heat pump system reduces the average electric power for cabin heating by 33% (3.51–2.34 kW) and increases the driving range by 12% (77–86 km) at 0 °C ambient temperature. A combination of the heat pump system with seat and panel heating ensuring same comfort as PTC heating, reduced the energy consumption for the heat up phase by 76% and increases the driving range by 32%. At +10 °C ambient temperature the average electric power for cabin heating is reduced by 63% (1.76–0.66 kW) and the driving range increased by 14% (91–104 km) by means of the heat pump system. Simulation results showed that the most efficient heating strategy for +10 °C is a combination of the heat pump system with seat heating including a reduction of the cabin temperature from 22 to 15 °C. This combination reduced the average electric power from 1.76 kW (PTC heating) to 0.43 kW (−76%) and increased the driving range from 91 to 107 km (+17%).

Acknowledgments This chapter was based on work conducted in the EU project "OPTEMUS", which has received funding from the European Union's Horizon 2020 research and innovation program under grant agreement No. 653288.

References

1. Farrington R, Rugh J (2000) Impact of vehicle air-conditioning on fuel economy, tailpipe emissions, and electric vehicle range. In: Earth technologies forum Washington (2000), NREL/CP-540-28960
2. Hao X, Wang H, Lin Z, Ouyang M (2020) Seasonal effects on electric vehicle energy consumption and driving range: a case study on personal, taxi, and ridesharing vehicles. J Clean Prod 249, ISSN 0959-6526
3. Fanger PO (1970) Thermal comfort: analysis and applications in environmental engineering

4. Pingel M, Backes D, Lichius T, Viehöfer J (2015) Akustische und thermische Komfortuntersuchung innovativer Heizsysteme. ATZ - Automobiltechnische Zeitschrift 117(11):44–49
5. Kaysser S, Leipe A, Ziegler H (2012) Elektrische Strahlungsheizung für ein Kraftfahrzeug und Verfahren zum Betreiben derselben. German Patent DE102012221116A1
6. Zhang Y, Wyon D, Fang L, Melikov A (2007) The influence of heated or cooled seats on the acceptable ambient temperature range. Ergonomics J
7. Schmidt C, Praster M, Wölki D, Wolf S, Treeck C (2013) Einfluss der Kontaktwärmeübertragung in Fahrzeugsitzen auf die thermische Behaglichkeit (FAT-261). Forschungsvereinigung Automobiltechnik e.V.
8. ISO 9920 (2007) Ergonomics of the thermal environment—Estimation of thermal insulation and water vapour resistance of a clothing ensemble
9. Parson K (2014) Human thermal environments, 3rd ed. CRC Press

Printed in the United States
By Bookmasters